Praise for
Wild About Horses

"Scanlan's discussion of horses through history—horses in war, horses in sports, wild horses, the gentling of horses through 'whispering'—at times approaches the lyrical and has the power to move the reader to tears." —*Booklist*

"A fascinating documentary record of our obsession with the horse. It is a work of great craftsmanship laid against a template of unconditional love."
—Michael Enright, *Globe and Mail*

"A journey of discovery. . . . As an author, journalist, producer, and editor, Scanlan's expertise is evident. *Wild About Horses* is an extensively researched visual and literary work that is interesting, informative, and entertaining."
—*Quill & Quire*

"Here is a book with horsepower!" —Rita Mae Brown

"In addition to exploring the history, mythology, and contemporary iconography of the horse, *Wild About Horses* is delightfully well written. I am flattered to be one of the writers / horse lovers to have been interviewed by the author."
—Maxine Kumin, *The New Yorker*

"Scanlan, a three-time National Magazine Award winner, knows how to gently poke fun at our horsy obsessions and also how to tease the horse manure from the many wonderful stories of horse sense. . . . The simple fact that horses have intruded upon our imaginations to such a vast extent suggests that our bond with the beast is more than merely practical, and Scanlan is an ideal guide to that secret world of connections, with its crazy and sublime turnings."

—*Kirkus Reviews*

"All those years of hanging around barns and show rings—and asking questions, always asking questions—have paid off. *Wild About Horses* is a very special and insightful book." —Ian Miller, two-time World Cup champion and co-author of *Riding High*

"Confirmed horse devotees, those who admire the equine animal from afar, and anyone who wants to learn about the variety and complexity of human-horse relationships will find this book indispensable."

—Elizabeth Atwood-Lawrence, author of *Hoofbeats and Society: Studies of Human-Horse Interactions*

"*Wild About Horses* will bring pleasure and a lot of new information to the old hand as well as the complete greenhorn. Scanlan reminds us that the horse is essential to who we are."
—Holly Menino, author of *Forward Motion: Horses, Humans, and the Competitive Enterprise*

"*Wild About Horses* is proof that our continued love affair with the horse is far from over."
—Jo-Ann Mapson, author of *Hank & Chloe, Blue Rodeo, Shadow Ranch, Loving Chloe,* and *The Wilder Sisters*

WILD ABOUT HORSES

Our Timeless Passion
for the Horse

LAWRENCE SCANLAN

Perennial

An Imprint of HarperCollins*Publishers*

A hardcover edition of this book was published in 1998 by
HarperCollins Publishers.

FIRST PERENNIAL EDITION PUBLISHED 2001.

Designed by Sharon Foster Design

Library of Congress Cataloging-in-Publication Data is available.

ISBN 0-06-093114-0

01 02 03 04 05 ❖/RRD 10 9 8 7 6 5 4 3 2 1

For my parents Bern and Clare,
and in memory of their parents,
Terry and Rose, Leonard and Gertrude.

Lawrence Scanlan on Radish, left, with Skip Ashley on Spot. Wyoming, 1997.

ACKNOWLEDGMENTS

Those who helped shape and inform and otherwise nudge this
book towards completion are too numerous to name
but I remain in their debt.

Special thanks to J. D. Carpenter, Cindy Fisher, Skip
Ashley and his wranglers in Lander, Wyoming, Carol Risbridger
and all who rode with me in the high desert. Thanks also to
Elizabeth Atwood Lawrence, Vicki Hearne, Maxine Kumin, Jan
Whitford, Christina Salavantis, Jim Elder, Dick and Adele Rockwell,
Vickie Rowlands, Barbara Whittome, Louise Dennys, Larry
Ashmead, Allison McCabe, Kathi Bayly, Scott Richardson, Sharon
Foster, Beverly Sotolov, and to horses named Luke and Radish and
Mabe. My editor, Sarah Davies, and my partner, Ulrike Bender,
both read the manuscript numerous times and I am
extremely grateful for their diligence and care.

Two prominent riders and trainers, one American, one
Canadian, taught me a great deal about the nature of the horse-
human bond in the course of my dealings with them. Monty
Roberts and Ian Millar encouraged me, each in his own way, to
try to fathom that bond. They believe, as I do, that however
intricate and elusive and ancient that connection is,
it is not one beyond words.

CONTENTS

INTRODUCTION

The best thing for the inside of a man
is the outside of a horse.
LORD PALMERSTON

IN JANUARY 1996, the Canadian magazine *Equinox* devoted twelve full pages to a piece called "Why Humans Love Horses." The writer had sought answers in history and mythology, sport, religion and literature. He had joined a cattle drive in the Albertan foothills to lend color and texture to his very personal exploration of the horse-human bond. In the end, he admitted that he, too, had a touch of horse fever.

The writer, I hasten to add, was me.

My aim then was to fathom that age-old love affair between humans and horses. But with the magazine article I had only begun to plumb the depths of the subject. While collaborating eight years earlier with the top-ranked Canadian show jumper Ian Millar on the writing of his memoirs, *Riding High*, I had begun walking the edge of the horse world. Like an ant treading the rim of a glass of water, I was never far removed from that world but never immersed in it, either. Mine seemed a useful vantage point — for several magazine pieces of the

long and horsy kind (including one on Spruce Meadows, that
Mecca for show jumping in Calgary, Alberta) and for a horse
biography, *Big Ben* (that horse, a two-time world champion
show jumper in the late 1980s, had dominated the sport into
the early 1990s). By the time I sat down to write "Why
Humans Love Horses," the ant was peering over the rim of the
glass and in danger of falling into the water. Or was it something headier, like wine?

My own horse fever I trace to my conviction that to write
of horses I had to ride horses. I had taken lessons at Wilmarny
Farm, the stable down the road from my rural Ontario home,
for four years. Long after it made practical sense, I kept hanging around horses. Riding infiltrated travel: every trip taken
or planned was an excuse to see new country from the back
of a horse — Spanish coast, Costa Rican rain forest, Peruvian
desert.

I became a capable enough rider, though perhaps I flatter
myself. During the Albertan cattle drive, I once found myself
riding somewhat sidesaddle when that was never my intent.
After the long day's ride, I slid my sore carcass down the side
of the horse — good form on an English saddle, bad form on
a western one — and hooked the front of my nylon jacket on
the pommel (the horn), which left me dangling helplessly a
foot off the ground, a vertical appendage stuck to the horse's
side. Lightning, the horse I was glued to, reversed, prompting
neighboring horses to ponder panic while I, oh hapless dude,
did the same. I called to the rancher, who was, happily, close
by. "Uh, Keith . . ." His large paws reached over and patiently
lifted me down, like a coat off a hook.

Wild About Horses allowed a more rewarding run at the
horse-human question, and the writing led me deeper and
deeper, pleasurably so, into the world of horses. The research
took me on more rides — in badlands Wyoming, in back-
woods Vermont, in high desert California — to gather stories,
impressions, opinions. Objective research gave way to personal journey. The horse made incursions into my brain: the

old railway path where I daily walk my dog, it dawned on me, would make a dandy place to ride. At the end of all this, I strongly suspected, stood a horse I would call my own.

My horse-inspired treks took me into the heads and barns of wise and passionate horse people; took me into the rich literature on horses, where scientists, soldiers, showjumpers, poets, cowboys and jockeys all muse on what it means to connect with a horse. The horse in war, in legend, literature and film; the horse in human history — I pored over this vast territory. I bought and borrowed scores of horse books and stacked them in little piles on the floor of my study (eventually, each book issued Post-it notes like thin yellow fruit).

I had come late to horses, had the convert's zeal. I went on collecting the life stories of horses and riders, putting, as it were, faces and tails to names. If I was even to think of calling myself a horseman I had to know the stories: of Phar Lap, Ruffian, Dick Francis, Bucephalus, Wild Horse Annie, A. F. Tschiffely, Comanche. I had in mind a book that would mark my own trail of discovery as I learned the well-versed-horse-person's vocabulary. For two years, I talked to horse people and gathered horse stories. Call me "The Man Who Listens to Horse Lore."

My research was shaped by my appetite for equestrian partnerships marked by intensity or eccentricity. I was fascinated by the notion of the epic ride, which led me to Aime Felix Tschiffely and his ten-thousand-mile journey on horseback in 1926. The plight of the wild horse led me to Wild Horse Annie, the woman who headed the charge to stop the slaughter. A minor obsession with trick ponies brought me to wonder horses named Marocco and Clever Hans. My horse whisperer file grew to include John Solomon Rarey, who took the horse world by storm in the nineteenth century the way Monty Roberts has the twentieth.

Out of all this came *Wild About Horses*, a book that considers our fascination with the horse from many angles. Clinical evidence and expert opinion help inform the book, but my investigation also happily took me into storytelling, where the truth is altogether more elusive, yet tidier, too.

Consider, for example, the case of Colonel, a sorrel draft horse who lived on a farm in Minnesota in the early 1900s. (Note that I say *who* lived, not *that* lived. I'll come back to this in the epilogue, but throughout the book I refer to a horse — and, indeed, to any animal — as *he* and *she*, not *it*.) In summer, children on the farm would hitch Colonel to a boxlike sled with room inside for about a dozen little ones. The huge horse with the blaze down his long nose would patiently stand for the harnessing, then turn to face the giggling crew, as if to say, "Everyone safely on board?" There were no reins — there was no need. The children would shout, "Let's go, Colonel!" He would pull them across a wooden bridge that spanned a creek, but sometimes — out of mischief? — he would haul them right into the creek, where he would stop for a long, cool drink while the children squealed with delight as the cold water flowed over their legs.

Colonel seemed to know when someone had pitched out. Perhaps he heard the sound of a tiny body tumbling into the grass; perhaps he felt the load lighten a touch. He would stop, his massive head would swing around to survey matters and he would proceed only when all were safely seated again. When Colonel had his fill of baby-sitting, he would head back to the barn, and no amount of cajoling from those in the drag box could dissuade him when his mind was made up.

It was the custom then to stable the horses all winter long. But in the spring, when the smell of new grass wafted in, the plow horses would grow restless. It became a ritual to release the horses all at once. Everyone on the farm would gather as the great wing gates to the pasture were opened and halters were slipped off the horses in the barn. Down by the open gate with the farm folk would stand Colonel — no mere horse, after all, but one of the family.

In the anticipation preceding one such ritual, no one noticed that four-year-old Eleanor was on the other side of the gate. The moment that the farmers slapped the flanks of the horses inside the barn to begin the stampede was the moment that Eleanor chose to cross in front of them. Everyone froze.

Colonel surged forward, bent one knee and knocked the little girl to the ground, then straddled the now screaming child and faced the herd. An instant later the other horses thundered round the great horse, the way water skirts a boulder in a brook, and galloped on to the pasture. Colonel leaned down and nuzzled the child, then stepped back as Eleanor's mother took her in her arms.

What I like about the story is how it captures the two towering aspects of horses: their danger (the draft horses would have trampled the girl, without malice, in their mad dash for new grass) and their generosity (the one huge horse created a living shelter).

The story of Colonel comes to me from a friend, who got it from a woman in New Mexico, who heard it firsthand from Marie, younger sister of little Eleanor. The story is like that lucky penny — to be found, put in horse-mad people's pockets and then passed on.

What to make of that story? Is it a story about the horse as guardian angel, anecdotal proof that human affection for horses goes the other way, too? Or would a scientist cast a cooler eye on the incident and put it down to instinct or herd response? The more I pondered humans and horses, the more questions I had.

Meanwhile, evidence from many quarters was telling me that interest in horses was more widespread and more intense than ever.

Book publishers certainly knew it. *The Horse Whisperer*, a first novel in 1995 by the British writer Nicholas Evans and inspired, in part, by the life of the American horse trainer and gentler Monty Roberts, was followed two years later by *The Man Who Listens to Horses*, the life story of the same Monty Roberts. (As luck would have it, I collaborated with Monty and contributed the introduction and epilogue to the book. To live for a time at his horse farm in California was an enriching experience and it took me deeper yet into the land and language of *Equus*.)

No one could have predicted that a novel about a horse gentler (the film, starring Robert Redford, came out in the spring of 1998) and a gentler's autobiography would grip audiences as they have. Where both books converge is in their insistence that the horse, or at least connecting with a horse, has healing power. Readers in great numbers seem interested in horses as sentient beings capable of memory, feeling, sorrow and joy.

Readers of modern literature have rediscovered the horse in the writing, too, of the American author Cormac McCarthy, whose career turned a corner in 1992 with the publication of the novel *All the Pretty Horses*. Set in outback Texas and Mexico in the early 1930s, this brilliant novel reads at times like a paean to horses. Along with its sequel, *The Crossing* (the second in a planned trilogy), it describes long journeys on horseback, and must have touched some deep chord, for it has won two major prizes and sold five hundred thousand copies — a staggering figure for a literary work.

When Elephants Weep: The Emotional Lives of Animals, published in 1995, seemed to usher in a raft of books — many of them about dogs — that celebrated animals as fellow creatures. Other contemporary books have argued for animal minds, animal souls and, of course, animal liberation. Convinced there is more here than meets the eye, we have grown increasingly fascinated with the inner lives of animals.

At the same time, a new wave of adults has come to embrace riding: in the past five years the Canadian Equestrian Federation has grown steadily each year from 8,255 members to more than 11,000 — evidence of a growing number of competitive riders. There exist more than 630,000 horses across Canada and some 78,000 riding establishments. In thirty-three countries around the world, 500,000 children belong to Pony Clubs and the number is growing.

In the United States, the ever-expanding horse industry contributes more than $15 billion to the economy. The numbers are staggering: 14,000 sanctioned horse shows annually, close to 300,000 young people involved in 4-H and pony programs or Pony Club. The number of horses in America has grown from 6.6 million to 6.9 million in the past year.

In many areas, demand for good riding horses has pushed up their price. Guest ranches all over North America are enjoying unprecedented business as "cappuccino cowboys" (or so their genuine counterparts sometimes call them) indulge their fantasies. There exist some five hundred dude ranches in America, and interest in both ranch vacations and in starting dude ranch operations is up dramatically.

Even in Hollywood horse fever is raging. Billy Crystal had never ridden a horse before starring in *City Slickers*, a thin comedy about urban men on a cattle drive; smitten, he bought the horse he rode in that film. Kiefer Sutherland starred in *The Cowboy Way* and later took two years off from acting to compete in rodeos. Robert Duvall in his other life is an accomplished show jumper.

From *Vanity Fair* to advertising, from children's literature to the boom in equine art, the horse is ubiquitous. There are more horses in North America now than there were in the 1800s when horses powered the family farm. At the turn of the century, horse numbers declined dramatically as the world mechanized, and that trend might reasonably have been expected to continue until the horse became little more than an ornamental species. But the trend did not continue. Today, there are 60 million horses around the world and the number is once again growing.

Horse fever still rages. For many of us, the attraction to horses amounts to an obsession.

My best answer to the question Why *do* humans, or some of us anyway, love horses? is the book you hold in your hands. In the Epilogue I will come back to that question. Perhaps by then both of us will be a little wiser and better informed about the great tribe of riders and horses, as well as what stirs the passion — "the right magnificent obsession," as the poet and horsewoman Maxine Kumin calls it — that draws us to them.

CHAPTER I

HEAVENLY HORSES

And, behold, a pale horse; and he that sat upon him,
his name was Death.
BOOK OF THE APOCALYPSE

We had no word for the strange animal we got
from the white man — the horse. So we called it
šunka wakan, *"holy dog."*
LAME DEER, SIOUX MEDICINE MAN, IN
Lame Deer: Seeker of Visions

"WHAT'S BRED IN the bone will not out of the flesh,"
a thirteenth-century proverb has it. In at least some
of us, the love of horses is indeed bred in the bone — an
ancestral seed seemingly passed on from generation to generation, as genuine as blood marrow, as clear to the eye as an
insect locked in amber.

In a family, even one living in the tangle of a city, far from
stables and pastures, a particular daughter or son may simply,

inexplicably, be born with a longing for horses. Poll any class-room, urban or rural, and ask children to name their favorite animal: bet on the horse coming out on top.

A horsewoman I know "rode" brooms as a toddler; Ian Millar, later a world champion equestrian, "rode" his piano bench as a boy. Growing up in Ottawa, Ontario, young Millar watched desperately for rent-a-ponies in the neighborhood and followed westerns like a hound on a trail. Maxine Kumin, the poet, seems also to have been born horse-mad, and as a child would give camp blankets and lumps of sugar to passing cart horses in her suburban Philadelphia neighborhood. She prayed for, "lobbied mightily for," a pony.

I stand in awe of that intrinsic drive and I wonder: Where does horse fever come from? How far back can we trace its roots? The literature on myths and legends suggests that our memory of horses is of a collective, almost universal, sort.

Customs among horse cultures were rich and varied, but the similarities were sometimes striking. Plains Indian tribes were true masters of the horse, and among the Crow, for ex-ample, horse and rider were so much considered as one entity that a warrior would strike a man's horse in the face to insult the owner. Is it pure coincidence that among the Siberian Kirghiz, a continent and centuries away, to strike another man's horse or even to speak harshly to the horse, was akin to insulting the owner?

Few societies have failed to be touched in some way by the horse. Early humans formed horse cults, created complex cos-mologies with winged horses, explained the rising and setting of the sun as the work of heavenly horses pulling the orb across the sky. All over the ancient world, the horse figured almost as much in human consciousness as the sun itself.

It was believed that in the afterlife horses, too, would be resurrected, some wearing the gold-plated girths, richly embroidered saddle cloths and bronze tail rings they had been buried with, standing in their graves as if ready for one last, glorious ride.

In Cormac McCarthy's *All the Pretty Horses*, a veteran of cavalry wars opines on the nature of horses — how "the souls of horses mirror the souls of men more closely than men suppose," how horses love war and have no need of heaven. And when the old Mexican is asked what would happen to horses' souls were horses to disappear from the face of the earth, he assures that "God would not permit such a thing."

Such confidence in the future of horses can only come from exceedingly deep roots — the oral traditions and ancient mythologies that were the precursors to our literature. There, the horse loomed large.

Always were, McCarthy's sage seems to say of horses, always will be.

It was long thought that the first rider mounted a horse some four thousand years ago, but recent discoveries keep pushing that moment further and further back into the recesses of time.

During the 1960s, a horse-head carving — with engraved lines that could be a halter — was found in southwest France in a cave known as La Marche. The carving is up to fifteen thousand years old. "Upper Palaeolithic people," says paleontologist Paul Bahn, "were of exactly the same intelligence as we are. You'd expect that it would dawn on them that they might be able to do more with horses than simply throw a spear at them when they were feeling hungry." The image of Ice Age people "galloping across the chilly grasslands of Europe" might conflict with our preconceptions, says his colleague Richard Leakey, but may well be accurate.

And what to make of thirty-thousand-year-old horse teeth that show evidence of crib biting — the habit of biting on hard objects that only corralled or tethered horses engage in? Did confined horses serve as pets, as decoys, or as a ready source of fresh meat? And when did that first courageous human ride the first horse? Teeth, it turns out, may offer the best evidence for dating the elusive moment when *Homo erectus* became *Homo equestris*.

In December, 1991, *Scientific American* featured an article by an American-Russian team of two anthropologists and one archeologist who argued from their own field research on the steppes of Russia that humans rode horses at least six thousand years ago. Like forensic scientists probing dental records to identify a murder victim, the scientists scrutinized, even X-rayed, horse teeth found in a burial site.

Over time, horses accustomed to wearing a bit sustain subtle damage to their teeth, quite apart from natural occlusion. The two kinds of dental erosion are clearly distinguishable under a scanning electron microscope. The scientists were thus able to buttress their claim that riding predates even the invention of the wheel by at least five hundred years.

As described in the *Scientific American* article, excavation at the burial site uncovered a seven-year-old stallion who bore telltale marks on his teeth: the beveling, the scarring of the enamel, the cracks and their location, all supported the conclusion that the stallion had been ridden. The stallion was surely not the first horse to be ridden, but he is the first horse *known* to have been ridden. His remains, the authors enthused, therefore constituted a "most spectacular find."

The team uncovered more than two thousand horse bones at the site — a village called Dereivka, near Kiev. These Copper Age people, known as the Sredni Stog, clearly had eaten horsemeat. Equally apparent was that the horse had special meaning for them: the aforementioned stallion's head and left foreleg had been ritually deployed to mark a sacred site.

The magazine piece also featured a photograph recreating what the assemblage of the stallion's bones might have looked like. It is a haunting image of a ritual that was widely observed in pre-Christian Europe. The skulls of two horses are mounted on poles, with other poles forming a rough tripod. From the uppermost and nearly horizontal poles hang the horses' hides, ghostly skins flapping in the wind.

"The assemblage," the authors wrote, "recalls Indo-European myths of a horse that bears souls to the gates of Hades,

1.1   For eons, the ancients marked sacred sites by mounting horse skulls on poles.

where dogs stand guard." The mood of the photograph is unsettling: the skulls seem to possess a dark vitality. The positioning of one head just below and ahead of the other, suggests motion, as if the creatures had indeed embarked on some terrible journey to the underworld.

The scientists found near the stallion's bones those of two dogs, along with perforated pieces of antler — possibly the cheek pieces of a bit — plus a clay figurine shaped like a boar and fragments of other figurines in human form.

The ritual mounting of horse heads on poles, the authors added, "was conducted well into this century among the Buryat and Oirot peoples, who live between the Altai Mountains and Lake Baikal in Soviet Asia; it may persist there to this day."

I find it intriguing, even gratifying, that such rituals endure despite the apparent global domination of Western culture. In one part of the planet, girls in riding helmets trot their horses around an arena on a Saturday morning while an instructor reminds them politely, "Heels down, girls. Heels down." Elsewhere on the globe a horse is beheaded and skinned to create a sacred sculpture whose origins date back sixty centuries or more.

I was amazed at how many disparate cultures in the ancient world used horse bones either to curse the living or to honor and equip the dead.

The Vikings, for example, believed that to put a curse on an enemy you simply beheaded a horse, set the skull on a pole and pointed it toward the intended victim. Keeping the skull's mouth agape was thought to enhance the curse.

The thirteenth-century tomb of Sir Robert de Shurland at Minster in the British Isles suggests that perhaps the Vikings were right about horse skulls and curses: the bones have power. Near the tomb is a sculpture of a horse's head emerging from the sea; on the church tower close by is a weather cock in the shape of a horse's head. Why all this horse symbolism?

Legend has it that Sir Robert — for whatever reason — buried a priest alive. When the king happened to be sailing nearby, the knight swam his horse two miles out to sea to beg, or purchase, his pardon. Which was forthcoming. Back on the beach, an old woman warned the knight that the extraordinary horse whose marathon swim had just saved him would also cost him his life. Sir Robert took his sword and lopped off the horse's head. And that, it seemed, was that. But a year later the churlish knight was walking on that same beach and chanced across his horse's skull, which he kicked in contempt. A bone splinter entered Sir Robert's foot, and the infection killed him.

But one image recurs in the prehistory of the world — the horse skull strategically mounted to mark a grave, in the manner of the Sredni Stog. The Patagonian Indians of South America would mark the death of a chief by sacrificing four horses, stuffing their skins and propping them up on sticks — precisely as was done in Copper Age Ukraine.

In the middle of the thirteenth century, as part of a diplomatic mission from the pope, a portly Franciscan monk named Johannes de Pian del Carpine spent some time in the court of the Great Khan. The aged friar took note of Mongol burial customs.

"The dead man," he wrote, "is laid in the earth with his tent, and provisions and mare's milk beside him. A mare with foal at foot and a saddled and bridled gelding are buried with him. Then they eat another horse, stuff the hide with straw and prop it up on two or four poles, so that in the next world the dead man shall have somewhere to live, a mare to provide milk, a foal to start another herd, and a horse to ride."

In fourteenth-century China, an emperor killed in battle was entombed with four young female slaves and six guards, and a vast mound built over them; four horses were then made to gallop around the mound until exhausted. The horses were finally killed and impaled on poles, there to stand guard and serve the emperor in the afterlife. In later eras, tiny terra cotta figures of horses and soldiers stood symbolically in their stead.

Ancient people often buried their dead with a horse. The Scythians, who lived by the Black Sea, would strangle up to fifty of a dead king's finest horses — along with members of the royal guard — to keep him company in his sprawling tomb. I cannot argue with their logic: the dead, after all, needed transportation in the next world.

In West Africa, it was common to bury a man wrapped in

a shroud made from the skin of his favorite horse. Hungarians beheaded the dead man's horse and buried the skull and cannon bones alongside the body of his deceased rider. The horse's hide was stuffed with hay and laid beside the man.

Lithuanians in early Christian times created separate cemeteries for their horses, the animals sometimes buried standing and fully tacked. Some Prussian tribes cremated the deceased on horseback — again, so the departed soul could ride to his reward in heaven.

The fate, then, of many ancient horses was to be buried in the earth alongside their riders, but the horse had another role, too — as an oracle, a link to the heavens. Before battle, ancient Slavic peoples dedicated a white war horse to their god of war and if the horse walked cleanly through a line of crossed spears without catching his feet, prospects for the engagement were deemed good.

The ancients loved their horses but feared their gods, and so the animals the ancients most prized were sacrificed to appease the heavens. This, too, is a bloody refrain in the common history of horses and humans.

One South American tribe in Patagonia would mark the birth of a boy in this fashion: a colt or a mare was secured by rope and cut from the neck down to the heart, which was then removed. The infant was thrust into the gore of the still-throbbing chest cavity to ensure that the newborn would one day make a fine horseman.

In the early days of the empire, the Romans also practiced horse sacrifice. A winning horse in a chariot race would be stabbed with a spear by a priest. Homage was thus paid to Mars, god of war and harvests. That horse's blood was used to purify livestock; the tail was brought to the ruler's house to ensure his prosperity and that of the empire. The horse's head was nailed to a wall, then decorated with a string of loaves to ensure a generous harvest.

The ancients of India and of Greece, torn between their

strong feeling for the horse and the greater tug of sacrifice, dared not spill equine blood. Instead, they would select white steeds and drive them into the sea, where they drowned.

The Irish and their highland cousins either refused — even in pagan days — to eat horseflesh, viewing it as the worst sort of taboo, or they made ceremonial meals of horse stew. A medieval Welsh chronicler records in rather grim detail how the northern Irish baptized their new kings in horse blood:

> The whole people of that country being gathered in one place, a white mare is led into the midst of them, and he who is to be inaugurated . . . comes before the people on all-fours, confessing himself a beast with no less impudence than imprudence. The mare being immediately killed and cut in pieces and boiled, a bath is prepared for him from the broth.

1.2   In myth, legend and literature, the white horse — especially the white stallion — figures prominently.

Sitting in this he eats of the flesh which is brought to him, the people standing round and partaking of it also. He is also required to drink of the broth in which he is bathed, not drawing it in any vessel, nor even in his hand, but lapping it with his mouth. These unrighteous rites being duly accomplished, his royal authority and dominion are ratified.

As Oldfield Howey points out in his book *The Horse in Magic and Myth*, the partaking of the body and blood of the horse has a Catholic sense of sacrament about it. Like the white Eucharist, the pure white horse confers divine grace.

Horses Make a Landscape Look More Beautiful, a book of poetry by the American writer Alice Walker, takes its title from the words of a Sioux medicine man: "We had no word for the strange animal we got from the white man — the horse. So we called it *šunka wakan*, 'holy dog.' For bringing us the horse we could almost forgive you for bringing us whiskey. Horses make a landscape look more beautiful." If the horse was a *holy* dog, then implicit is the belief that the horse came from heaven.

The horse, of course, came out of a Spanish galleon. The first one stepped off on January 2, 1494, when Columbus landed on North American shores for the second time. But as the conquistadors began to enslave the indigenous people of what is now Latin America, their greatest ally, it seemed, was the horse. The vastly outnumbered Spaniards, naturally, encouraged the belief that horses did indeed come from heaven, that they were gods.

The Zempoaltecas, a tribe in sixteenth-century Mexico, concluded that only bridles stopped the Spaniards' horses from eating their human masters and anyone else within reach. When the horses neighed, the Indians rushed to feed them, hoping to placate them. And thus we come to the story of Morzillo, Hernando Cortés's proud black horse.

During a skirmish in the early 1500s, Morzillo took an

arrow in the mouth; Cortés, one in the hand. Numerous accounts of horses in war describe how furiously they sometimes attack the enemy with their teeth and hooves, and this was how Morzillo reacted to his pain. The Indians panicked at his onslaught.

Later, Morzillo — healed from the arrow — was left in the care of Cortés's Indian allies, who said they could treat a splinter in his foot. Still in awe of the horse, they named him Tziminchac, god of thunder and lightning. They decorated him with flowers and brought him what they took to be delicacies — cooked chickens. Perhaps Morzillo pined for his master or perhaps the divine diet was his undoing. In any case, the horse wasted away and died.

A century later two Spanish priests arrived in that same place, the first Europeans since Cortés. They were astonished by what they saw — a temple large enough to accommodate a thousand people, and in it an enormous statue of a horse seated on his haunches. This was the legacy of El Morzillo, the horse who became a god.

Stolen, escaped or abandoned, the horses brought by the Spaniards would rapidly spread north to the plains, where people there created their own myths and stories to explain this apparent gift of the gods. A Piegan woman who died in 1880 at the age of 116 told such a story to Montana pioneers. She was called Sikey-kio, Black Bear, and this is her account:

One evening a daughter of the chief saw a bright star and muttered, "If that star were a young man, I would marry him." Next day a fine young man appeared to her on the trail and took her to his star world.

Later, a glimpse of her village far below saddened her, so her husband made a long rope of buffalo hides and gently lowered her back. She bore a son, but later died and an uncle cared for the boy.

Up above, the great chief saw how destitute were the people down below. He went to the boy — his grandson —

and asked him to bring wet clay, which grew and finally moved. Then the great chief called a council of the trees, birds and animals, for he was their ruler.

"I have made a horse for my grandson," he told them, "an animal for him to ride and carry his burdens. Now each of you will give me of your wisdom, that this horse may be perfect." The pine tree contributed a tail, the fir tree a mane. The turtle offered hoofs, the elk his great size, the cottonwood fashioned a saddle. From the buffalo, wolf and snake came hair bridle, fur robe and straps.

The boy then mounted the horse and rode back to his people, who were both astonished and envious of the colts that followed. When the boy became a young man, he told his uncle he was leaving forever but not before turning every fish in a nearby lake into a horse. The people were to catch the horses as they emerged from the water. "But you, uncle," he said, "are to catch my old mare. Catch her and her alone."

When the young man rode the mare into the lake, the water bubbled and issued hundreds of horses. Many were captured but others escaped and formed wild herds.

Last out was the feeble old mare, and though the people mocked him, the uncle dutifully tied her to a stake by his lodge. That night, as the moon rose over the hills, the old mare neighed three times and thousands of colts and fillies galloped from the woods and surrounded the man's lodge. That is how the great chief gave horses to the Piegans.

Every horse people on the North American continent told such legends to explain the coming of the horse, an event that had reshaped their lives as hunters and traders and warriors. Before riding into battle, for example, the Navajo warrior would whisper in the ear of his horse. He was really talking to himself, so much did he feel at one with his horse. "Be brave," he would tell his horse, "and nothing will happen. We will come back safely."

What despair the Navajo must have felt when the Creator,

who gave them horses, quite suddenly took them away. Late in the 1800s, the U.S. Army crushed the last resistance of Indian people. Take away their horses, the army realized, and you destroy their spirit of defiance. The horseless Navajo, therefore, were confined to Fort Sumner in New Mexico. There they went on conducting their religious rites, including one called the Enemy Way, long performed on horseback. Determined to maintain these rituals, they improvised.

The Navajo decorated long sticks and painted them to represent the different colors of horses; they stuck crude horse heads on one end and then "rode" these sticks in their ceremony as the children of pioneers rode hobby horses.

The Navajo rode their stick horses, they said, so that one day when they were free — for surely the Creator would not make them live so wretchedly for long — their children would once more have horses.

Like the Piegan, pioneers would also engage in myth making. Across the prairie and decades apart, settlers in covered wagons, drovers and trappers, all claimed to have seen a wild white stallion of heroic dimensions. In *Moby Dick*, Herman Melville mentions the horse — "a magnificent milk-white charger, large-eyed, small-headed, bluff-chested" — and countless other writers of the day did the same.

J. Frank Dobie, in his book *On the Open Range*, describes an event said to have occurred in Texas around 1848. German colonists had tied a little girl called Gretchen onto the back of an old mare, who then wandered off in search of grass. The mythic white mustang appeared and led the errant mare back to his herd; he cut the ropes that held Gretchen in place and lifted her up by the collar, in the way that mother cats transport their kittens.

Later, after the girl had rested, the white stallion put her on the mare again (the story ducks the detail of retying Gretchen's saddle knots) and instructed the mare to return her to her family camp. The little girl, eloquent for one so young, described the horse as "pacing with all the fire of a mustang emperor" and noted that "something about him" kept her calm.

Several accounts describe how mustangers — or wild horse traders — tried in every way to catch the white stallion. One story has a hundred men on their best mounts trapping him in a circular gully and chasing him by turns, until each mounted horse is exhausted. Toying with his enemy, the white mustang scales a cliff and disappears over the rise.

By the late nineteenth century, a small fortune was being offered for his capture. Near the Rio Grande, he was finally caught by three vaqueros, who all roped him simultaneously. They staked the noble gray to a spot by water and grass, but he refused any of it and after ten days lay down and died.

Similar accounts describe newly broken mustangs appearing to make a choice between liberty and death. In one case, a stallion immersed his head in water up to his eyes and steadfastly refused to come up for air despite desperate hauling on the reins by three men.

For centuries, the most important horses in civilization were mythic horses, horses who were imagined into existence and given a place among the gods. The ancient Greeks, for example, created an immensely complex cosmology in which horses — both mortal and divine — played a part.

Poseidon, god of the sea, created the horse as part of a contest: whoever contrived the most useful object for humankind would win the honor of naming a new city. Poseidon struck the ground with his trident and a splendid white stallion sprang forth. The gods assembled were indeed impressed with this new creature, and more so as Poseidon explained its various uses, and especially in war.

Athena, goddess of war but also skilled in the arts of peace and diplomacy, struck the ground with her bronze spear and offered what at first seemed a laughable creation — the olive tree.

"You have both done well," said Zeus, god of the sky.

"Both are great gifts, but in days to come mortals will use one well and the other badly. They will use the horse for war and make it into an instrument of destruction. But the olive tree will give them food, oil, wood, comfort and prosperity." Athena named the new city after herself.

The pairing of human and horse, according to Aesop, the Greek writer of fables, came about when humankind went to the aid of the equine kind. Aesop tells of a horse and a stag living harmoniously in a lush valley. But one day the stag announced that he wanted all the grass for himself and drove the horse away with his sharp antlers. When asked, a man who chanced by said he would help win back the grazing ground — if the horse in turn helped him. The man returned with a bridle, mounted the horse and drove the stag off with a spear.

The horse was delirious with joy and asked the man, kindly, to remove the bit so he could get on with his grazing. The man would have none of it, for he realized with sudden clarity what the strong, sleek horse had to offer. The horse had struck a contract with the human, and their fates would be forever entwined.

In a league of his own among mythic horses was Pegasus, the winged one. A constellation would be named after him and Shakespeare would describe him as "pure air and fire." Pegasus is with us yet. A Hollywood studio has adopted as its symbol the winged horse.

Pegasus's father was Poseidon; his mother, the Gorgon Medusa. When drops of blood from her severed head fell into the sea foam, Poseidon used them to create a winged steed. Snowy white and immortal, Pegasus was born with unimaginable speed. Apollo and the nine Muses (the goddesses of music, poetry and the arts) rode him often. But no mortal had ever ridden Pegasus. The first to do so was Bellerophon, whom the gods had asked to slay the Chimera, a grotesque fire-breathing monster that was a lion fore and a goat aft.

Bellerophon accomplished that task; emboldened, he then sought to ride Pegasus to Olympus and join the gods. Like

1.3 Mortal on immortal: Bellerophon on Pegasus, the winged horse of Greek mythology.

Icarus, who flew too close to the sun, proud Bellerophon was brought back to earth. The enraged Zeus sent a gadfly to sting Pegasus and the great horse bucked off his rider. Blinded by his injuries, Bellerophon never rode him again.

From Greek history, we remember another horse — the famous wooden horse of the Trojan War who never moved a muscle yet succeeded in destroying a city. After a ten-year-long siege, Ulysses despaired of ever taking Troy. But before departing, he consulted a soothsayer who advised him to construct an effigy of a mare — big enough to conceal soldiers inside. Ulysses did so, then told the Trojans that the great wooden horse he put before their city gates was a gift to placate Athena and help him get safely home. In fact, Ulysses sailed only as far as the nearby island of Tenedos. There, he waited for a signal from spies planted near Troy.

1.4   Greeks hauling their wooden offering: a gift horse the Trojans should have looked in the mouth.

The siege-weary Trojans cheered the apparent departure of the Greeks. One elder suggested that the horse be brought into the city, but the priest Laocoon warned his fellow citizens against it in Virgil's *Aeneid*.

Somewhat is sure designed by fraud or force—
Trust not their presents, nor admit the horse.

The Trojans did just that. During the night, Greek soldiers inside poured out, opened the city gates and set Troy on fire. The Trojan horse fell into legend and became an enduring lesson in human folly.

Another refrain in mythology links horses and healing. Greek mythology, for example, features Chiron the Centaur, a wise creature half human and half horse who educated the heroes of Greek legends and taught medicine. The prescription for those suffering from wounds and disease? Ride a horse to cheer your spirits.

Similarly, a tradition in Scandinavian folk lore links healing with an incantation about riding a horse. The legend refers to Odin, the Norse god of war who also taught horse handling, and his invincible son Baldur the Beautiful. One day Baldur's horse slipped and broke a leg but was miraculously cured when Odin did two things. He tied a black thread with seven knots and then recited this verse (later slightly altered by Christians, who retained its essence):

Baldur rode. The foal slade [slipped]
He lighted, and he righted,
Set joint to joint, bone to bone,
Sinew to sinew,
Heal in Odin's name.

Many folk tales ascribe magical powers to old, lame, sick and misshapen horses. Such stories were told in eastern Europe for centuries, and archeologists recently discovered why. They excavated a thousand-year-old cemetery in the Hungarian town of Keszthely and discovered the remains of crippled and malformed horses buried in elaborate graves. Most of these horses had an extra tooth or bones that showed evidence of lameness.

In Ireland, some folk believed (perhaps some do yet) that a pure white horse confers upon its rider the gift of wisdom in curing physical ills. Thus, a family with a sick child might in days gone by have stationed one member out on the road to wait for a rider on a white horse. The rider would be asked to stop and prescribe a cure — chicken soup, porridge, black beer.

Parts of the horse's anatomy were thought by various cultures to possess power, sometimes healing power. "Fortune," an Arab adage goes, "is attached to the horse's mane." Horse skulls, or so eleventh-century Hungarians believed, brought health, wealth and happiness, and when affixed to the interior walls of their houses, protection against evil spirits. Ancient Chinese doctors gave powdered horse heart to patients inclined to forgetfulness. The Spanish fashioned amulets made from stag horn and black mare's tail to ward off evil. Huns drank horse blood to seal an oath; Mongols used horse shoulder blades to tell fortunes. And in many cultures through time, it was widely thought that dreaming of horses brought the dreamer good luck.

Dreams defy interpretation, but many books on symbols and images — along with so-called "dream encyclopedias" — associate dreams of horses with sexual desire. We still speak of "unbridled" passions; Carl Jung linked dreams of horses to lust and fecundity. The ride is still about letting go.

The late Paul Shepard, an American ecologist and the author of *The Others: How Animals Made Us Human*, argued in that book that our ancestors "had 'horse' in their stomachs and in their heads in every sense . . . of all the animals the horse may well have been the most elusive and the most intelligent, the one deepest in human dreams and imagination . . ."

"It is no wonder," he wrote, "that we love them."

Mythological roots go deep, but it is also true that many of the religions that followed these so-called pagan systems similarly embraced the horse. Job in the Old Testament was fulsome in his praise of the horse: "The glory of his nostrils is terrible. He paweth in the valley and rejoiceth in his strength: he goeth on to meet the armed men. He mocketh at fear and is not affrighted; neither turneth he back from the sword."

Muslims, if anything, felt an even greater affection for the horse than did the Hebrews. The desert Arab was known to bring his horse, who was considered part of the family, into his tent. "The greatest of all blessings," the prophet Muhammad once said, "is an intelligent woman or a prolific mare."

Berbers saw the horse as the holiest of all animals, with the greatest concentration of *baraka* or grace — the black horse most of all. It was believed that by throwing yourself between the legs of a horse who had been to Mecca you could command the protection of the owner. In desert lands, a horse stable was seen as a sanctuary, much as a church or monastery would later be in Europe.

In Haitian and African voodoo, the possessed human — who snorts and paws the ground — becomes a horse to be mounted and ridden by the spirit.

Christianity adopted the horse in its own way. The horse may be the only animal whose consumption by Christians was prohibited by papal decree. Pope Gregory II wrote to Boniface: "Thou hast allowed a few to eat the flesh of wild horses, and many to eat the flesh of tame ones. From now on, holy brother, permit this on no account."

The edict likely reflected love and respect for the horse but also turf wars between Christianity and pagan religions. If the horse was central to pagan rites, what better Christian strategy than to declare horseflesh *verboten*? Besides, war was almost a constant and so was the need for cavalry horses.

Christians, too, had stories to celebrate the horse, and the saints to go with them. St. Dunstan, a bishop in early England, who must also have worked as a farrier, once saw the devil in a horse he was shoeing. He made Satan promise never to enter or disturb a building in which a horseshoe had been hung —

hung U-shaped, of course, lest the good luck all run out. Even Admiral Nelson nailed a horseshoe to the mast of his ship *Victory* in the belief that it offered protection.

St. Anthony, a saint often associated with horses, was apparently being assailed by a mounted Egyptian king when the normally placid horse turned on his master, bucked him off and delivered a fatal bite. Even a pagan horse knew a saint when he saw one.

The people of the Russian steppes, likely the first horse people, saw the next world as a mirror world to this one — with everything backward. This may explain why even today, state and military funerals often feature an ornate riderless horse with boots fixed in the stirrups and the toes pointing backward. In 1963, the funeral cortege of President John F. Kennedy featured a black "caparisoned" horse (called, ironically, Black Jack), his stirrups holding empty black boots.

For the funeral in 1997 of Diana, princess of Wales, six black horses of the King's Troop, Royal Horse Artillery, pulled her coffin on a gun carriage the two miles from Kensington Palace to Westminster Abbey. It looked stately, but the ride was terrifying for the three troopers, each on the left horse of paired Irish draft horses.

In such processions each ornately uniformed rider carries a white-corded whip, which he will lay against his forehead as a silent signal to the horse to slow, stop or turn. The horses are noted for both their discipline and their spirit, and only the most gifted, confident riders dare mount horses used to pulling one-and-a-half-ton gun carriages in bold crisscrossing maneuvers where speeds may exceed thirty miles an hour.

On the day of Diana's funeral the risk of spooked horses was huge because of all those flashing cameras, the great rain of mourners' flowers tumbling onto the coffin, and the fact that the lead mounted officer suffered from an acute allergy: even one sting from the many wasps buzzing around the flowers might have cost him his life and induced untold calamity.

1.5   Albrecht Dürer's *Four Horsemen of the Apocalypse*: biblical symbols
      of death, war, famine and conquest.

The troopers (who all sighed with relief at the end) were par-
ticipating in an age-old tradition — using horses to honor the
dead. The plumed black horses who pulled funeral coaches
over cobbled streets in Dickens's time also have a place in a
worn tapestry full of corpses and horses.

———

If the horse mattered near the beginning of human existence (as the cave paintings at Lascaux in France suggest), so may he matter at the end.

The Hindu deity Vishnu, apparently, has descended into the world nine times in various animal forms. The tenth visit, at the nadir of our depravity, will mark the end of the world. Prophecies paint Vishnu in the sky, seated on yet another white and winged horse, a kind of Pegasus of doom. The horse is poised with one hoof in the air: the moment it descends, the earth will cease to exist.

There is ample time yet, one hopes, before the hoof comes down — or before the biblical Four Horsemen of the Apocalypse (death on the pale horse, war on the red, famine on the black and conquest on the white) ride into final view. In the meantime, the horse will likely remain a powerful force in our collective unconscious.

R. B. Cunninghame Graham, a noted horseman who died in 1936, had his own ideas of paradise. "God forbid," he once wrote in a letter to Theodore Roosevelt, "that I should go to any heaven in which there are no horses."

CHAPTER 2

WILD ABOUT WILD
HORSES

*Halted in animated expectancy or running in
abandoned freedom, the mustang was the most beautiful,
the most spirited and the most inspiriting creature
ever to print foot on the grasses of America.*
J. FRANK DOBIE, *The Mustangs*

IN THE SUMMER of 1997 I went to the Great Divide in
search of wild horses. In a way not much happened that
week of riding in Wyoming, and my bare-bones report could
read as follows.

The skies every day were blue and vast; the silence and
sense of wilderness near absolute. It was June in the high
desert and the wildflowers were stunningly in bloom. I felt
saddle-sore much of the time: our guide, a lean, copper-
skinned man of Cherokee ancestry, pushed us as hard as he
dared. The fourteen riders — from the U.S., Britain, Holland

and Germany, with me the lone Canadian — camped in chalk-colored canvas tepees in the desert two nights, in the mountains for three. All week long I rode a sixteen-hand roan (horse height is measured in hands, a hand being four inches, so a sixteen-three horse is 16 hands three inches). The horse's name was Radish, as in horseradish. Only later did I discover that the state flower of Wyoming, a ubiquitous red cone flower called the Indian paintbrush, tastes like radish. That detail seemed, finally, to form a tidy circle around one of the better weeks in my life. And yes, we saw wild horses.

This was no expedition where perfect strangers bond for survival's sake. I have been on whitewater canoe trips near James Bay in July when eleven consecutive days of cold rain, the odd three-mile portage and the threat of hypothermia forced paddlers together. In Wyoming, we dined surprisingly regally, stayed warm and dry, had the sun and the stars for company.

I took a photograph that speaks to the mood of the trip. Our horses are behind the photographer, grazing with loose reins, as we often let them, but no horse is in my lens sight. We have stopped for a break on a windswept height of land. Their backs to the camera, ten mostly hatted riders have found perches on boulders. They sit alone or in groups of two or three, but the striking thing is how all face the same direction. None appears to be in conversation. All are looking out to an immense expanse of sky and rock and treeless plain, as if paying homage to the land. We sit as we might in a cathedral, or as if we had all read the Irish poet Seamus Heaney's line about landscape as sacrament, that it was to be read and pondered as text.

In truth, something *did* happen out there. Two riders from Vermont, Fran F. and Nancy O., fell off their horses. Fran did so in a calamitous Jane sort of way while galloping. The horse shied at something on a narrow trail, instantly converting all that forward momentum into lateral momentum in the way that horses do, leaving the rider suspended in the air.

Somehow Fran drifted down and landed on her feet, even kept her cowboy hat on, as if she had practiced this feat for the amusement of onlookers. She escaped unhurt. But I was right behind her and the what-ifs perhaps gave me more pause than they did her, for the calm never left her face. What if her foot had caught in the stirrup when that Quarter Horse bolted as he did? What if Radish had run her over? What if she had landed on her head, and not with feet-first elegance?

Nancy was grounded by an innocent-looking little buck that followed a leap over an equally innocent-looking ditch. She went right over the Arab's neck, somersaulted neatly and landed on her bottom with a thud. She had the look of someone awoken in a strange place with no idea of how she got there.

Our guide, Skip Ashley, quickly got off his horse and grabbed Nancy's: a matter of safety, he explained around the campfire that night, for if the horse had spooked, Nancy would have been vulnerable as she sat dazed in the sand. It looked to us at first glance that Skip, who had to endure our gibes, had instinctively declared his priority: horse first, Nancy second. Her only injury, it seemed, was a sizable splinter where bottom so abruptly met sagebrush. Nancy is a spare, muscled woman, a marathoner, and she shook off the injury the way she shook off Skip's suggestion that she change horses the next day. She got back in the saddle and we all rode off.

Only later, on the plane home, did she hear an odd scraping sound from her rib cage and feel a fist of pain when she moved a certain way. A ballpoint pen makes that sound when you click it. Nancy tried not to move that certain way, tried to duck the fists, and was shocked when the X-rays showed three broken ribs. "I am either one tough bird," she told me later, "or I so desperately valued a week of riding in Wyoming that I buried the pain."

Those were the few events that made the week what you might call uneventful. But when I later wrote the other riders to elicit their responses to the experience of riding and camping in the Wyoming outback, I got back the most extraordinary missives, longer and more detailed than most moderns have the time or inclination to compose. By then we had all had a few weeks to let the ride sink in, and clearly, it had made a deep and lasting impression — a bit like the prairie itself, which still bears the imprint in places of pioneers' covered wagons.

That whole week will remain fixed forever in my brain. The memory of it is a film I keep rewinding, starting now in the middle, now at the beginning, now at the end. When I tell people about running with the wild horses, some get it instantly. Others nod, but obviously, they do not speak the language of horses, or at least of horse lovers; do not grasp what it has meant to me to cross the Red Desert Basin in Skip's wake. To sit with my back against a rock high on a crest and feel nothing but gratitude. To ride bareback in my underwear — me squealing like a boy — as my horse swims across a pond, then circling back to do it again and again. To rejoice in the sight of mustangs grazing in silhouette on a far rise, catching, as I was, the last rays of the sun. To breathe and dream sage, so much that a whiff of it now calls back Wyoming. To look down from the saddle to the lupines and fleabane and Indian paintbrush and to feel such a sense of wonder. To feel whatever woes I claim to shoulder just slipping away, like the wind that took the dust of our horses.

Time and Space and a Horse are what we had out there. For the first time in my life, I felt I was seeing the world from the back of a horse. Another century had me in its grip. My last night in the mountains I hiked up to the ridge overlooking the camp and felt powerful emotions surging in me. (Or was I just out of breath?) I kept thinking about the Shoshone and how they built tepees overlooking the plains. We saw what they saw. Maybe they felt what pilgrims to Chartres or Mecca felt: an overwhelming sense of their own smallness. I rather like that feeling and I felt it the whole time I was out in the badlands, which were not so bad at all.

David Dary is the Kansas-born author of *The Buffalo Book* and *Cowboy Culture*, both of which convey the nineteenth-century west as it really was. In the latter book, he wonders why the flat, treeless plain "left an unmistakable feeling that man was rather insignificant. There was and is the feeling of an ever-present supreme being or force not always found in the woodlands and certainly not in the cities . . . Why does the vastness, silence, and solitude of the plains affect man as it does? It is a question that may never be answered."

The Canadian author Sandra Birdsell, in a much admired short-story collection called *The Two-Headed Calf*, renders the quiet of the prairie — "a quiet so intense and full," says one of her characters, "that he used to think he could scoop it out of the air and cup it in his hands."

It is what I missed most when I went home: that deep quiet that seemed to propel me, all at once, inside and outside myself.

A remarkable little book, *The Solace of Open Spaces* by Gretel Ehrlich, who ended up living and working as a sheep-herder and ranch hand in Wyoming, made sense of that lunar landscape, how it can be so comforting and discomfiting at the same time. Ehrlich writes of sleeping on the ground curled like a dog against the wind. (I remember doing the same, napping at noon under sagebrush on the sun-burned plain.) She writes of a weather front "pulling the huge sky over me," like a blanket, I suppose.

Ehrlich also mentions a widow on one desolate ranch who used to bring her saddle horse into her living room for company on winter nights. I did find solace riding Wyoming's open spaces, and I knew, all of us on that trip knew, you could not go just once.

Implicit in the letters was the notion that riding across the plains and over the mountains was a journey or pilgrimage. Something *did* happen on that trip. The letters formed a chorus to that effect.

Carol R. of Connecticut: "Riding through this terrain and experiencing these spectacular 360-degree views seemed to awaken all my senses. I wanted to learn about the wildflowers, to feel the earth, dig for fossils, drink that ol' cowboy coffee at breakfast, listen to the coyotes at night, watch the gorgeous sunsets and the rising moon. I wanted to do it all, without constraints."

Whatever touched Carol about the experience was rooted in the marriage of horse and landscape. Her horse, that landscape. Free, for a while at least, of all encumbrance. The riding both animated her and gave her a deep sense of calm. Long hours in the saddle acted like a salve, bestowed a sense of tranquillity, gave her time to think about life, to see what mattered, what made her content.

Carol had come on what Skip calls the "wild horse–mountain ride" with her seventy-something father. Both had been around horses all their lives. But what she had thought might be an enriching father-daughter experience — a sharing of their common love for horses — became something far more. She did not intend for the trip to be in any way spiritual, but her horse, a dun pony named Armajo, quite literally took her to a place she had never been before.

A journey on horseback will sometimes do that. The horse, first literally and then in a deeper way, slows you down and lets you take note of land and sky. E. T. Hall in his book on human perceptions of time, *The Dance of Life*, once moved a small remuda of horses in southwestern America several hundred miles and was struck by the impact. It was nothing like covering the same distance in a car on a paved highway: "The horse, the country, and the weather set the pace; we were in the grip of nature, with little control of the rate of progress . . . it took a minimum of three days to adjust to the tempo and the more leisurely rhythm of the horse's walking gait . . . I became part of the country again and my whole psyche changed."

Several years ago, a Canadian Broadcasting Corporation

producer named Kathleen Flaherty needed a holiday. She was tired and overworked; her beloved cat had just died. Since she had always loved horses — "from afar" — she booked a weekend ride on the Skyline Trail in northern Alberta's Jasper National Park. Flaherty liked it so much she went again a year later, this time with pounds of recording equipment.

Her hour-long radio documentary is a striking mix of commentary and texture — horses clipping stones while crossing mountain brooks, horses whinnying to one another, the sound of horses being tacked up, the crackle of campfires and the banter they induce. Like Carol, Flaherty talks about the joys of cowboy coffee, about the vivifying effect of spending time on a horse in such grandeur.

The slow pace of a walking horse (in her case a bay mustang named Busley), the solitude, the quiet, all work their magic. "I realize I am living exactly in this moment," Flaherty is heard to say. "My mind finally slows down enough to deepen — just like my breathing." She wonders why escape induces such euphoria, and answers her own question: "Because there is time and space to allow for contemplation and wonder."

The documentary ends fittingly with Joni Mitchell's "Night Ride Home" and its haunting lyrics about the pure sweet power of escape.

The strangers who gathered that first Saturday evening at Skip's ranch near Lander, Wyoming, to meet the horses and one another all felt much the same as Flaherty did a week later. Connecting with a horse on the bald, empty prairie seemed to cleanse us.

Carol's father, John B., the oldest rider on the trip, lives in rural New York State. "When I am with a horse," he wrote, "my mind becomes a complete blank other than the thought of caring for him — all my worries (if I have any) are dispatched as soon as I am with him." He called the trip "a horseman's dream" and he was amazed by the soundness, stamina and endurance of the Quarter Horses we rode. He struggled to put into words what it all meant.

Tracy T., a young English rider and clearly one of the most capable equestrians among us (the reward for her skills was to be assigned a feisty Thoroughbred-Quarter Horse cross), was similarly reeling from the experience weeks later. I had sent her a copy of the photograph described earlier, which she cherished, for it "somehow managed to capture the sense of us all being overwhelmed by our surroundings."

Tracy had typed her note on what looked like sky — cottony clouds adrift on light-blue paper: "I still can't tell you why Wyoming now has such a grip on me, but there is just something magical out there. Everything is tougher — the horses, the weather, the terrain."

Kathy C. of Connecticut wrote in her own hand a five-page letter that zeroed in nicely, I thought, on why we were all so touched. It had a lot to do, she believed, with the view — the view from a saddle: "The sense of the vastness of the land, the beauty you could never see zooming along in a car, the wildflowers and rocks, are all so special when seen from the back of a horse." Kathy had been going west on riding trips for twelve years and she always came back feeling renewed, as if she had been to a monastery.

Sally B.'s letter was emblazoned at the top with the computer-generated silhouette of a large black horse and, in bold lettering, "Let's do it!" This was the phrase that Skip ritually and loudly announced at the end of a break or the beginning of the day when the horses were all saddled and the riders all set. The emphasis was on *do it!* He could have said, "Let's move out!" or "Saddle up!" However, it was never anything but "Let's *do it!*"

"Oh, that week in Wyoming," wrote Sally from home in Vermont. "It is with me all the time."

Fran, like Sally, wonders if in another life she was a cowpoke or an Indian. She, too, was hard pressed to find the words for that week in Wyoming defined by the horse: "Only those who were there know what that means."

On day one, the Sunday, we would occasionally crest a ridge and see small bands of wild horses in the next valley. No surprise, for there are some 450 mustangs roaming free in the Great Divide Basin. But we never got closer than a few hundred yards before the band's stallion began to pace and prance and the herd soon vanished.

I had never seen someone sit a horse as Skip did. Watching him roll cigarettes in the saddle, I imagined he might with equal grace and balance write a letter there, thread a needle, whittle a carving. He started off riding his imperious senior stallion, Mr. T. — called that because of his T-shaped blaze. But the old boy was not up to it, and when Skip switched over to a young paint horse named Spot, Mr. T., big black blue-eyed Mr. T., followed along. Or rather, led. Imposing in every way, he knew the trail as well as Skip did and had the respect, even deference, of every horse in our number.

Skip wore brown leather chaps and boots with spurs that jangled when he walked. I rarely saw him with boots and spurs off, so below the waist he seemed unduly thick, and the jangling round spurs, ribbed like a gear, made him sound metallic on the ground. In the saddle and above the waist was another matter. His shirts were always open and cut at the shoulder; oftentimes he rode shirtless under the desert sun. The bone necklace, the small blue tattoo of linked eagle feathers ringing one bicep, the headband he sometimes wore, Apache-style hair: all created an effect. On that paint horse, he needed only a lance and leggings to take any observer into the eighteenth century when his ancestors hunted on the plains.

The son of a Cherokee father and a Czech mother, he playfully called himself a "CheroCzech." But it was his Indianness, not his Europeanness, that emanated from him on that trip. I had the sense of being challenged and watched over at the same time.

Funny the things I remember about the journey: Skip's instruction to me and the others, that first day on the high desert, not to crowd him. How unusually frank he was and how (also unusually) I took no offense by it. How I amused myself on climbs by responding to Radish's blowing with some of my own. How wet his red flanks were at the end of those climbs and how quickly, even miraculously, he dried and recovered at the top. Radish up to his belly in a watering hole happily pounding the surface with one hoof, like a kid scissor-kicking at the end of a dock. The stark beauty of a dead oak that presided over the mountain camp. Skip's admonition one time *not* to break any branches when we passed through a grove of oaks. The *way* he said it put me to mind of Gandalf the wizard counseling Bilbo Baggins and a gaggle of dwarves in *The Hobbit* as they entered an enchanted forest. I felt, and resisted, a great compulsion to do just what Skip said we should not.

I remember the rivers — the Little Popo Agie, the Wind, the Sulphur, the Sweetwater, Crooks Creek. The towns nearby — Shoshone, Arapahoe, Lost Cabin. The geography and "points of interest" from my map of Wyoming — Split Rock, Three Crossings, the Antelope Hills.

On that Monday, near the end of a long hard gallop I was right behind Skip as he veered to dip into a twenty-foot ravine, not slowing in the slightest. As he descended he turned and gave me this smile that conveyed a snapshot of his own character: the mischief, the joy, the daring and the dare. So you think you can ride with me, do you?

After lunch that day, Skip suggested that those who were sore and wanted a lighter ride could head back with Bruce, one of the wranglers, because those who rode on with him would be a lot sorer at the end of the day. He guaranteed it. Along with nine others, I opted for (more) soreness. Some riders later confided to me that though they had been around horses all their lives, never had they ridden as hard or as fast as they did that day. Skip reckoned we covered forty-five miles.

The gallops were sometimes ten or fifteen minutes long, and it was during one of these later in the day that it happened. For safety's sake we rode cavalry-style, two abreast, up the two tracks of a dirt road through the desert. I kept at or near the lead, because Radish liked to run and because his rider disliked the taste of dust. A ridge rose on our right, then flattened out as we continued the long canter, but it was only when we neared the edge of the ridge that we saw them. Twenty-five mustangs matching us stride for stride and not more than fifty yards to our right.

Who knows how long they had been galloping unseen on the other side of that ridge? Our own horses had no doubt sensed them, and what we had taken for high spirit was really a keenness to run with their wild brothers and sisters. If that whole week was a film in my head, this was a moment for the highlight reel. Unreal was more like it. We let out whoops; we gasped.

Gill E., the Suffolk mother of two who had come to Wyoming with her sister Ro, remembered the gallop with the wild horses as "the most powerful moment in my life — primitive even. We were all part of one herd, pounding up the track with Skip as our leader . . . across that incredible landscape."

Led by a black stallion, with many chestnut and paint mares and several foals, the horses could have veered off at any time. They did not. We rode side by side for a few miles, then they put on a burst, crossed the road ahead of us and ran on our left for a few more miles before finally disappearing over a hillside. How sleek they were. Skip told us that in the spring, many wild horses are thin and mangy. But not in June. After generous spring rains, the coulees still held water and the grass remained lush in the valleys. With their long manes and tails flying, these horses looked impressively healthy — as if a groom had fussed over them all morning. But no human had ever touched these horses.

Riding alongside a herd of domestic horses would have been a thrill in itself. These, though, were wild horses, free horses, and that lay at the heart of our euphoria. "I think," wrote J. Frank Dobie in *The Mustangs*, "that wild horses have more

2.1   Young filly grazing: the wild horse remains a powerful symbol of
freedom and high spirits.

secrets than gentle ones." The wild horse still symbolizes freedom and high spirit, and always has. Pioneer accounts, penned
during wagon train treks west, point to how common it was
to see free-roaming horses. Yet familiarity did *not* breed con-

tempt. Trappers and travelers, settlers and soldiers, thrilled to see them as we thrilled to see them a century or so later.

After admiring a fine sorrel stallion one day, Matt Field wrote in 1879 from the Santa Fe Trail that "a domestic horse will ever lack that magic and indescribable charm that beams like a halo around the simple name of freedom . . . He was free, and we loved him for the very possession of that liberty we longed to take from him."

Some of us had flown halfway around the world for the privilege of gazing upon free horses, and we took joy and sustenance from their playful little dash across the desert.

Once common on the plains, wild horses are now relatively rare, and that, too, elevated the moment. According to various estimates, two to five million mustangs (a derivation of the Spanish word *mestenos*, meaning wild or untamed horses) roamed the West in the early 1800s. Travelers reported seeing herds so massive that their movement across the horizon continued unabated from dawn to dusk. But by the end of that century, the wagon trains had carved deep ruts in the Santa Fe Trail and myriad other paths west. The taming of the West meant the shrinking of lands available to the free horse. Still, their numbers would occasionally get a boost: the defeat of Indian nations in the 1870s meant that a host of war horses and buffalo runners were taken from their tribal owners and released onto the plains. And during the Great Depression of the 1930s, drought and hardship hit hard at farmers. Huge numbers left their dusty acreages and simply abandoned their horse herds, letting them loose on the prairie. But nothing could augment wild-horse numbers. Other forces were at work. The horses were free, all right. Free for the taking.

Hundreds of thousands of wild horses were rounded up to serve in the Boer War. Industry had need of them to make chicken feed, fertilizer and hides. Ranchers took what wild horses they needed as cow ponies but resented the rest competing with their cows for grass on public lands. Ranchers, game managers, men in planes — all shot wild horses.

Hope Ryden, whose book *America's Last Wild Horses* helped stem the flow of horse blood, asked several of those flyers why they would venture into such inhospitable places to gun down such hardy horses. It was not as if the horses existed in overwhelming numbers. By this time — the late 1960s — the mustang population in the United States had dwindled to a paltry seventeen thousand and some observers were predicting that the wild horse would go the way of the passenger pigeon and the dodo. The horse hunters had only one explanation for their actions: the horses were "no good."

Some scientists unwittingly blessed the mass culling by noting, rightly, that horses roaming the plains were descendants of once domesticated horses. (Though not always: researchers in the early 1990s determined that horses in the Pryor Mountain Wild Horse Range, on the Montana-Wyoming border, bear strong genetic links to the Spanish horses brought by Columbus.) Free-roaming horses were deemed to be not true wild horses, therefore, but mere *feral* horses. Those who love horses prefer the word *wild* and hate the word *feral*. Same horse, different word. Yet the world turns on words.

The common language used to describe wild horses seemed to warrant their slaughter. Cattlemen called them "jugheads" or "broomtails" and maligned their conformation. An entry in *The Encyclopedia of the Horse* does not remember them fondly: "By the nineteenth century the typical mustang tended to be hammer-headed, ewe-necked, mutton-withered, roach-backed, cow-hocked and tied-in below the knee. These defects were generally ignored by artists, but cruelly displayed in early photographs."

Some old photographs seemed even to exaggerate these conformation faults. One arresting nineteenth-century photo in a book of the 1950s, *The Indian and the Horse*, shows a somewhat shaggy, angular pony, and on his back a two-year-old boy wearing a feathered and horned headdress that gives him the look of a medicine boy-man. Below the photo is a quote from an observer of the time who was less than im-

pressed with Indian ponies: "A typical product of the indis-
criminate coupling and winter hardships of the prairie horses —
small, tough, deer-legged, big-barreled, with slanting quarters,
mulish hocks, a hide fantastically flared and blotched with
white, and one wicked glass eye that showed the latent devil in
his heart." Whether the quote refers specifically to the photo is
unclear, but they seem a perfect match. The pony does indeed
possess a glassy eye.

Even today, some cattlemen in the Chilcotin Range of
British Columbia call wild horses "garbage" — "undesirable
and breeding in the wild, like starlings."

And yet the wild horse possessed undeniable attributes.
The Plains Indians rode only captured or stolen mustangs,
who won praise in some historic accounts for their speed and
intelligence, their courage in the buffalo hunt, their iron-hard
hooves and their uncommon freedom from foot and leg problems. Other observers commented on how docile the mustangs
were (perhaps owing something to how they were introduced
to the concept of riding by their Indian owners). They were so
easily managed, said one white traveler who tried an Indian
pony, that they made the Indian rider look better than he
actually was. But hammer-headed or not, the several million
ponies of the plains were almost wiped out.

To understand what happened to all those horses and who
saved them from extinction, you have to know something
about Velma Bronn Johnston.

She was born in Reno, Nevada, in 1912, the eldest of four
children. At the age of eleven, she contracted polio and spent
six months in a body cast, which left her able to walk but her
body misaligned. Pain and fatigue would plague her all her
life. Some of her schoolmates taunted her for her disability,
but she found solace in poetry and drawing and in the animals
on her parents' ranch. She knew what it was to be an outcast.

One morning in 1950, Johnston was driving along Highway 395 to work in Reno when a truck hauling horses cut in
front of her car. Shocked by a stream of blood dripping from

the truck, she followed the van to a slaughterhouse and watched from behind bushes as a yearling was trampled to a pulp between terrified stallions. The horses had buckshot wounds; some stood on bloody stumps after their hooves had worn off from running over rocks; one stallion had had his eyes shot out. What Johnston saw and heard that day would change her life. She would take on the U.S. federal government, specifically the Bureau of Land Management (BLM), which was encharged with managing wild-horse populations. Her most bitter opponent there, Dan Solari, would dub her "Wild Horse Annie," and the name stuck. Friends began to use it with affection.

Meanwhile, powerful enemies lined up against the wild horse. Cattle ranchers and sheep farmers complained that wild-horse herds ate their hay and competed with their livestock for precious water and grasslands, that stallions trampled fence lines and stole their mares.

2.2   Mustang roundup, Arizona, 1980: less than 200 years ago, two to five million horses roamed the plains.

Bureaucracy lined up against the horses. In 1919, the American government — pressed by ranchers — issued a bulletin that dealt in part with what it called "wild or worthless horses." Citing economic reasons (sheep and cattle fetched far more money than wild horses), the Department of the Interior outlined its plan for "ridding the range of these worthless horses." Australia, meanwhile, declared its own intention to rid the Outback of "brumbies" — "a very weed among animals."

But the loudest death knell for wild-horse herds was rung by the burgeoning pet food industry. "Nuisance" horses were rounded up and slaughtered, and the meat canned for consumption by cats and dogs. It was not just the fact of the slaughter that appalled Wild Horse Annie but its scale and practice. Mustangers (horse hunters) had been using planes since the 1930s, and by the 1950s they had it down to a science, deploying aircraft equipped with sirens.

"The mustangs," wrote Johnston, "are driven at breakneck speed by planes from their meager refuge in the rough and barren rimrock onto flatlands or dry lake beds. There the chase is taken up by hunters standing on fast-moving pickup trucks . . . after a run of fifteen to twenty miles, [the horses], many of them carrying bullet wounds inflicted to make them run the faster, are easy victims for ropers."

The wild horse was then lassoed, with a heavy truck tire at the other end of the rope serving as anchor. By then, Johnston wrote, the frantic horse's sides were heaving and blood ran from his nostrils. The mustanger tied the horse's feet and the creature was pulled up a rough plank ramp (which often stripped the hide from his flanks) onto a stock truck where the ropes were removed and he was prodded to his feet.

Another option, graphically portrayed in a film from 1961 called *The Misfits*, starring Clark Gable and Marilyn Monroe, was to rope mustangs from a flatbed truck, hogtie them and then leave them in that abject state all night, to be picked up by another truck in the morning.

Terrified and often hurt, the horse faced a long journey

to an abattoir that might be many hours away, perhaps in another state or even in another country (horses were sometimes shipped to Canada). Left behind to die were young colts or horses too badly injured to load. Thus to slaughter went

2.3 After repeatedly trying to vault a fence in Montana, this mustang stallion got what he cherished: his freedom.

what Velma Johnston called "the harassed, abused remnants of our Western heritage."

The cruelty was methodical, even ingenious. In *America's Last Wild Horses*, Hope Ryden describes how mustangers in Wyoming's Red Desert — precisely where I rode in 1997 — would capture and gentle a mare, then sew her nostrils almost completely shut with rawhide or barbed wire. Unable to take in a full breath of air, her speed was reduced and she would act as a brake on the herd in the spring when another roundup occurred. Another braking tactic was to bend a horseshoe around the ankle of a mare, which bruised her when she ran. Only a quality mare was selected, wrote Ryden — "so she could perform double duty by bearing a good colt for the mustangers during the year she was free on the desert."

The ten-year battle to save wild horses from virtual extinction had Velma Johnston right in the thick of it. The one photo I have seen of her shows her in a beehive hairdo and oversized sunglasses, standing outside a corral. She might have passed as a visitor to a dude ranch; she was, by all accounts, a quietly determined woman whose house, with its horse-inspired paintings, ceramics, lamps and bookends, reflected her love of the horse. Or at least *the idea* of the horse. For tiny Velma Johnston was, in fact, allergic to horses.

But before she died in 1977, Johnston proved to be a burr under the saddle of every cattleman, bureaucrat or pet food profiteer who stood between her and the horses she loved. Like many people who will never see a wild horse in their lives, she felt it important that free horses have a secure place on the planet.

Hope Ryden's book and Velma Johnston's campaign began to show results. An organization called WHOA (Wild Horse Organized Assistance), with Annie at the helm, undertook surveillance patrols and field studies; WHOA raised funds for the beleaguered wild horse as well as its profile. The *New York Times* carried a front-page story on the plight of mustangs. Other publications, including *National Geographic*, *Time* and

Reader's Digest, followed suit. The latter ran a footnote to its story and suggested that concerned readers write to members of Congress. They did: one senator got fourteen thousand letters in one week.

The same stirrings for the wild horse that compel today's advertisers to feature horses — and often free-running herds of horses — in ads for cars, lottery tickets, beer, doughnuts and clothing also ran deeply in 1970. But so did feeling the other way. A letter to the editor of the *Nevada State Journal* around this time predicted, ominously, that "Wild Horse Annie will be called Dead Horse Annie in a very few short years."

Nine years after Johnston first followed that slaughter-house truck, the U.S. Congress passed what became known as the Wild Horse Annie Law. Using motorized vehicles and polluting water holes to capture wild horses were both banned. Later, in 1971, mustang allies got another law passed: this one protected wild horses and burros from capture, harassment or death.

"Congress finds and declares," the legislation read, "that wild free-roaming horses and burros are living symbols of the historic and pioneer spirit of the West; that they contribute to the diversity of life forms within the Nation and enrich the lives of the American people." Protective ranges were declared in Montana, Nevada, Wyoming and several other states. One, the Book Cliffs Refuge in Colorado, was dedicated in memory of Wild Horse Annie.

Today, the International Society for the Protection of Mustangs and Burros, which Velma Johnston helped found, oversees an adoption program, as does the Bureau of Land Management itself. The government occasionally culls wild horse herds to control their numbers, but members of the public are invited to "adopt" these horses. In principle, it is a good idea that often works well. But too many of the horses still end up at meatpacking plants, often because the owner lacks the considerable skill and patience required to gentle a wild horse.

The first wild horse I ever saw was also the most vigilant horse I had ever seen. Everything about him said he was not a horse to be trifled with. I was in California in March of 1997, in the high desert of the Cuyuma Valley, and the mustang was alone in a corral. With me was Monty Roberts, a horseman who had recently adopted this horse and another from the BLM and who planned to gentle one of them *in the wild* during the week to follow. A BBC-TV documentary film crew were to arrive shortly and record his success, or failure, and the entire chronicle would go into the book I was helping Monty write, *The Man Who Listens to Horses*.

While Monty attended to details of the shoot, I stood by the fence and watched this wild horse, a little bay. He would find a spot in the dusty corral as far from me as he could, then circle and eye me. His stare was unrelenting. Some years earlier I noticed a wolf in a field in northern Ontario, and despite the great distance between us, it was instantly and instinctively clear to me that here was no dog. The focus, the stare, the wariness, said so. Here, now, was no pack horse. And though he had been in the corral for some weeks and had surely seen his share of humans in that time, he remained highly suspicious of all two-legged types. Not once did he lower his head or bend one leg or show any sign of relaxing his vigil.

As I leaned on the fence and looked in, I thought of the words of a Scottish doctor named John Bell who observed wild horses on the China-Mongolia border in 1719. He had seen what we now call Przewalski's (pronounced pshuh-vahl-skeez) horse, chunky little pepper pots with black manes and zebra stripes on their back legs. "The most watchful creatures alive," he wrote. Dr. Bell might have said the same of this bay.

In a fifty-foot round pen, Monty typically takes thirty minutes to introduce a green horse to bridle, saddle and rider. But gentling a wild horse in the wild (it was actually a twelve-hundred-acre pasture — ample room indeed for a mustang to ignore a gentler's inquiries) drew deeply on Monty's reservoir of skill and patience. Monty followed him around that huge field, in walk, trot and canter, for thirty-five consecutive hours and for several days thereafter. The message to the wild horse was something like, I'm not going to hurt you, but you and I have to talk. And I'm not going away until you *do* talk to me.

I plotted the course of their deliberations for several days, sometimes looking down with binoculars from a high knoll. By the second day, the wild horse would let Monty approach on his mount, but only from the near side. Approaching from the right, or off side, was met with fierce resistance. Finally, after many hours, the horse did allow Monty to reach down and touch him, and that moment seemed portentous somehow. The cold wind up on that knoll seemed to ease; the sun got a little warmer. Man and horse were talking.

But even on day four, when the mustang — whom Monty had named Shy Boy — came to abide being haltered and even saddled, he struck Monty hard with a front hoof when the latter moved too quickly to untie a knot. Shy Boy also greeted the young rider chosen for that first ride with another blow.

2.4   Convicts play tag on wild horses in Wyoming: the games are part of the gentling process.

As wild as some of us are about wild horses, or at least *the idea* of wild horses, wild horses are clearly not wild about humans. Not at first, in any case.

Sometimes the task of gentling adopted mustangs falls to people who perhaps have yet to learn the wisdom of patience and trust — inmates in penitentiaries. At the end of the gentling process, horses from the correctional center in Susanville, California, for example, fetch $350 to $600 at public auction. It is not a lot of money for often a very good horse.

Tom Chenoweth, a veteran trainer who oversees that program, does not hold with the view that the wild horse is inherently a superior horse. "Some wild horses," he told me, "are exceptional horses, but that's true in any breed. Most people who come to the prison to buy a wild horse are after economy horses. They're generally caring, feeling people. They like the idea of a wild horse and they need a trail horse. And for that, the wild horse is in a class all its own."

The good ones are never rattled, for example, by stepping into a muddy bog. Where a domestic horse would panic and likely pull a muscle, Chenoweth notes, the wild horse calmly backs out. Of the 650 or so wild horses who have gone through his program, only two ever suffered from colic and none suffered navicular, a disease of the hoof that riders and trainers traditionally fear.

Tom Pogacnik, who oversees the U.S. government's wild-horse gentling program, similarly sings the mustangs' praises. "They have strong legs and perfect feet," he says, "and once they know you they're incredibly affectionate. They're not always pretty horses, but they're perfect horses. They're horses as nature designed them, not as humans have bred them."

In fact, one reason to preserve the wild horse is as a gene pool, to be dipped into as needed. As Karen Sussman (president of the organization founded by Wild Horse Annie) told me, "We're breeding pretty and beautiful horses, but pretty and beautiful may not last." Studies have determined that wild horses are more genetically diverse than any other breed of horse. Nature weeds out the weak, the lame, the unwary; those left are tough enough to endure harsh winters and summer drought and to resist disease.

Around the world, wild horses can be found throughout South America, southern Japan and China. France has its Camargues; Britain, the Exmoor, Dartmoor and New Forest ponies. The Australian brumby population, estimated to be as high as six hundred thousand, ebbs and flows but somehow thrives on one of the most unforgiving territories on the planet.

In Canada, few wild horses remain. The foothills of Alberta still shelter several hundred, and scattered bands have found refuge in northern Saskatchewan, the southern Yukon and central British Columbia. One place where they exist in significant numbers is on Sable Island, about sixty miles off the east coast of Nova Scotia.

A sandy spit some twenty-five miles long and less than a mile wide, the treeless island offers no shade in summer. The only food — coarse dune grasses, wild pea plants and low shrubs — is plentiful in summer, hard to find in winter. From winter gales, from bone-chilling fog, there is almost no shelter. Waves here can reach sixty feet high and the salt water washes far inland.

Various theories abound as to how horses came to be here. Did they swim ashore from a French ship that foundered in the 1600s? The "Graveyard of the Atlantic," they called the island. Many hundreds of ships and many thousands of sailors fell prey to its hidden sandbars and terrible wind — so many, in fact, that a rescue service was established on the island between 1801 and 1958. Did a lighthouse keeper import the horses as company? Did a Boston merchant bring them as food for shipwrecked sailors? The only certainty is that horses have been on Sable Island for hundreds of years.

During storms the horses huddle in circles, rumps to the incessant wind. Blown sand has reduced the windows of abandoned houses to frosted panes. You wonder how horses survive in such a hostile habitat. Some do not. Records of herd numbers date from 1891, when up to a hundred horses died. More recently, in 1958, half the herd of three hundred perished.

Moved by their plight, the Canadian government dropped bales of hay from planes, but the horses — apparently wary of the human scent that pervaded the bales — declined the food. A plan in the late 1950s to remove the survivors and sell them to the highest bidder was greeted with such outrage that then prime minister John Diefenbaker issued a proclamation. It read, in part: "The horses of Sable Island and their progeny will not be removed, but left unmolested to roam wild and free as has been their custom for centuries."

In 1979, a harsh winter again halved the herd to 150. Numbers are back up again but oil and natural gas discoveries nearby may finally be the horses' undoing.

One of the few remaining wild herds in the world untampered by human intervention, Sable Island horses are also — no coincidence — among the toughest horses in the world: sure-footed, fast and blessed with uncommon stamina. They have the look of Barbs, the chunky horses seen in North Africa. Their only enemy is nature, which fiercely and relentlessly culls the weak. I remember a photograph of an old, feeble stallion on the island, taking refuge during a summer storm behind what remained of an old building. That summer would be the last one for the old stallion.

He would leave a bold legacy. We should let nature continue its course on the island. One day on the mainland we may have need of such quality bloodlines.

But will the wild horses endure? And will the delicate ecosystems where they have found refuge last if the horses do?

One reason given for the roundup of wild horses on an artillery range in Alberta in 1994 was concern over a unique sandhills ecosystem. Similarly, wild horses on barrier islands off the eastern American coast are feared to be threatening delicate island ecosystems. Witness a coy headline from 1993 in the *New York Times*: "Is Misty of Chincoteague leveling dune grass?" New research shows that horses reduce the number and variety of plants, and with it the islands' defenses against erosion.

It is a tangled problem. If authorities let horse numbers increase, the horses will overwhelm the dunes and trample the grasses. The result could be "a sandbox with horses," as one critic put it. A geologist argued that "if a developer damaged these islands the way the horses do, he would be put in jail." On the other hand, island horses have many admirers. Those who have read *Misty of Chincoteague* as children often go as adults to watch the annual ritual: the roundup by firefighters, the culled horses swimming the channel, the auction that Marguerite Henry's novel made famous.

It is important that both horses *and* island grasslands endure. The sight of wild horses on land that is more or less theirs is an illusion, but one that fosters hope.

"Never look a gift horse in the mouth," the saying goes. Accept charity with grace and without question. I thought of that saying recently when I chanced across an unpublished article by Susan Blair Seward. She described how she came to own a mustang called Utah, a fiery little horse rounded up for adoption by the BLM and later taken to a horse shelter in Leesburg, Virginia, and from there to Seward's home in Tidewater in 1993. Utah's first owner could no longer afford to keep the horse, but she was an experienced horsewoman who knew mustangs. She had treated him with kindness and gentleness and gained his trust. I liked how Seward put it: "Utah fell into a pudding."

Seward had always been something of a horse snob, used to riding sleek Thoroughbred hunters. Her friends dismissed mustangs, much as they scorned the alley cat and the pound mutt in favor of purebreds with papers and haughty bloodlines. But something about rugged little Utah touched Seward, something beyond words. Though saddling him in the beginning took three brave people, somehow this woman knew that trust lay just around the corner. She was right. Seward

2.5   Though diplomacy typically governs horse society, these mustang stallions fighting over a harem are savage in battle.

never regretted buying Utah and she was struck by his toughness. She had to smile when her friends' pedigreed mounts succumbed to foot problems, breathing disorders and other common ailments unknown to Utah and his wild kin.

Seward wrote with eloquence: "There is a law of genetics which, in the hands of God and nature, is practised with unerring certainty: only the strong survive. Utah's flint-hard hooves, alert, wide-set eyes and iron constitution are no accident; they are the creation of the perfect by the divine."

Sometimes Seward laments the fate of Utah, a creature born in freedom. A little sadly, she takes out a map of the state of Utah and traces with her finger the tiny town of Milford, where the mustang caught his last glimpse of home — "his big world of hills, rocks and sky."

Still, Seward thought, this horse had become her partner and friend. During one gallop across a great field, with the whistling wind and the pounding of his hooves the only sounds, Seward felt as though she were flying. She remembered a passage in the Koran on Allah's creation of the horse:

"He picked up the golden sand and cast it to the wind, saying, 'Thou shalt fly without wings.'"

It seems terribly important to some of us that the wild horse endure. Richard Adams, the author of *Traveller* and *Watership Down*, once put it rather bluntly. "For our own sakes," he wrote, "we all need to feel and understand the value to the world of the wild horses, those paragons of natural power, grace and beauty. I will go further: we cannot afford to do without them."

Jay F. Kirkpatrick, a wildlife biologist who has spent twenty-three years of his life observing wild horse herds all over the world, including the Red Desert of Wyoming, would perhaps agree with both Seward and Adams. This man is also a specialist in reproductive physiology and his work on what looks to be a safe and efficient method of birth control for the herds may hold the key to their continued survival. He is a scientist, but one not immune to the inherent beauty of horses in the wild.

In a book called *Into the Wind*, Kirkpatrick writes movingly of what it has meant to him to watch mustangs grazing in the Pryor Mountains, or brumbies gathering at the billabongs of Australia, or a young mare emerging with a foal after hiding in the Book Cliffs of Utah.

> To see a stallion and a mare affectionately groom each other, amid the man-height sagebrush of Wyoming's silent Red Desert, fills me with awe. And, if there is anything more beautiful on this earth than standing alone in an Assateague marsh in a cold March sunset, watching a band of brown and white pintos graze peacefully while snow geese glide overhead and sika deer gracefully prance through on their way to some favorite feeding place, I want to discover it before my own journey is finished. I am richer, beyond description, because of these events.
>
> And *that* is the value of the wild horse.

CHAPTER 3

THE HORSE THROUGH
THE LOOKING GLASS

*For 98 percent of the last 6,000 years, the
horse was the fastest vehicle available to man.
It is no wonder it represents an ancient
archetype for combined power and speed.*

HAROLD B. BARCLAY,
The Role of the Horse in Man's Culture

To UNDERSTAND THE unique and powerful kinship that
humans feel with horses, we must look past mythology
to history — where the answers to large questions often lie in
any case.

The age-old chronicle of the horse is not *just* about the
horse in war, though a strong case could be made along those
lines (and thus chapter 5 is devoted entirely to the war horse).
The horse through time has served us in innumerable ways:
horse meat kept our ancestors' bellies full, horse dung fueled

3.1 Circus performer May Wirth (who seems to be head over heels in love with horses) on Joe, 1915.

their fires, horsepower pulled their wagons across continents, cleared land and pulled plows. The horse's spirit granted the animal a place among the gods, and the horse's regal bearing put *Equus* in the funeral processions of rulers and at the coronation of kings and queens. The horse's speed thrilled and entertained us and even helped deliver the mail; the horse's exquisite form shaped our notions of beauty. For millennia, poets and warriors, priests and commoners, took sustenance and inspiration from the horse.

As I probed the chronology of the horse, I was struck by the pure breadth of it: fifty-five million years of equine history. By comparison, the six thousand or so years of human partnership with the horse, and even the one million years of existence that *Homo sapiens* can lay claim to, look more like a wink in time. I was also astonished by how little humans have truly understood the horse, a fact best appreciated when you consider what our ancestors mistakenly fed the horse through the ages.

Another message I took from the pages of history is the notion that horse and human were somehow meant to be partnered: certain physical and social aspects pointed the horse in the direction of humans. Some would call it fate.

"There is nothing new under the sun," the saying goes, though its kernel of truth has come to be buried. Synonymous these days with world weariness, the phrase derives from the Book of Ecclesiastes. Wise King Solomon warns against the ephemeral pleasures and vanities of the world in one of the Bible's most poetic sections. (Song writer Pete Seeger some decades ago adapted Solomon's words as lyrics: "To everything, turn, turn, turn, there is a season . . . a time to kill, a time to heal . . . a time to laugh, a time to weep . . .")

In the world of horses, it might also seem that under the sun there can be nothing new. Or at least nothing more to be written. One estimate suggests that since the time of the Greek horse trainer Xenophon, a few hundred years before the birth of Christ, some forty thousand books have been written about the horse. Likely at least that many instructional guides, anthologies and horse biographies have also found their way into print. More recently, the surging interest in horses has spawned yet another round of books (and CDs and videos) on how to ride them, groom them, house them, shoe and show them, stretch them, heal them, breed them, select, train and touch them, feed and fathom them, jump them, even how to pamper them.

Through time, the great outpouring of words helped promulgate some bizarre notions. The Romans, for example, believed that mares, more than any other animal, loved their offspring. Almost two thousand years ago, the poet Pliny the Elder wrote that "a love poison called horse frenzy is found in the forehead of horses at birth, the size of a dried fig, black in color." This the mare eats after dropping her foal, he warned, and if anyone takes it, she will refuse to suckle the foal and the thief will descend into madness.

Through much of history (famously, perversely) wrong information on horses has been passed on, or new gaffes invented. The proof of the pudding, as it were, lies in what we have fed these creatures we claim to know and admire.

Consider, for example, the equine high life as envisioned by another Roman of Pliny's time, a certain demented emperor named Caligula. He was properly known as Gaius Caesar Augustus Germanicus (A.D. 12–41), and only behind his back did Romans of the day call him Caligula (it means "Little Boots," for he often wore *caligae*, or soldier's boots). His very name now conjures the decline of the Roman Empire.

Keener on games and horses than on managing the realm, Caligula built a huge amphitheater and filled it with water so that mock naval battles could be held as entertainments. He was like a little boy playing war in his bathtub, but he quickly squandered his fabulous wealth on this epic tub and other toys. At first generous, he later banished or murdered his relatives (save his sister Drusilla, whom he may have bedded) and for his amusement would have enemies tortured and slain before him as he dined.

He was also mad about horses, in particular one chariot horse — Incitatus, meaning "spurred on." Caligula made him a citizen of the Roman Empire, gave him a gem-studded collar, had a horse blanket sewn for him of royal purple (the emperor's color) and insisted that his oats be dipped in gold.

In a stable fashioned of marble eighteen servants waited on Incitatus. They fed him, or tried to, the things that Caligula imagined his precious horse would enjoy — mice dipped in butter, raw mussels, marinated squid and, of course, those golden oats.

Worse was yet to come. Caligula promoted the horse from citizen to priest and held a banquet to mark the occasion. Distinguished senators arrived to find Incitatus sitting on his haunches and wearing his jewels and a white bib. But when a platter of roast chicken was presented to the horse, Incitatus spooked, the table crashed and the banqueters fled.

Caligula had himself declared a god and was soon assassinated — the fate of many, but especially loony, rulers. Incitatus, erstwhile citizen and cleric, reverted to mere horse and thereafter was spared gilded oats, oily mice and chicken cacciatore.

Food for horses, it seems, is food for thought. Horses in ancient India were fed peas boiled in sugar and butter, root vegetables steeped in honey and, on military campaigns, wine as a sedative. Romans fed their horses a supplement of sparrow eggs. In the Algerian oases, horses ate dates and learned to spit out the pits. When water was scarce, horses were sometimes fed camel milk.

Legend has it that the Japanese hero Yokoyama Shogen fed his mount Onikage chopped-up humans. In medieval times in northern and western Europe, bread was offered horses when fodder ran scarce; beer, curry and oysters have all been fed to horses. Two seventeenth-century French noblemen would feed their horses three hundred eggs each before a race, convinced the yolks gave their mounts an advantage in speed. The Tungus, who live in Manchuria, may yet believe that the horse is at heart a carnivore. They start young horses on salted fish and move on to raw meat.

The brutal game of *bagai*, or *buzkashi* — a kind of equestrian free-for-all in which the object is to maintain possession, at whatever cost, of a goat carcass — was played by Mongol riders in the days of Genghis Khan. Today, a somewhat more refined version of the game is played in Afghanistan, where it has become the national sport. Full-time *buzkashi* players employ their own servants and grooms. Horses for this sport get strict training and exercise and are fed the usual alfalfa, oats, hay — and sometimes in winter, eggs and sheep fat.

Today, Western trainers strive to give their horses an edge, sometimes by adjusting the feed. Some trainers (like those *buzkashi* grooms dropping sheep fat in the food bin) give horses extra fat in the belief that it offers an anaerobic boost. Another Tungus-like tactic is to feed the horse creatine, a substance found in the juices of flesh. If this all turns out to be a mistake, we can simply add gristle and blood juice to history's already long and fatuous equine-menu.

Finally, certain trainers have taken to feeding their horses up to a pound of bicarbonate of soda just before a race by dropping a tube through the nose and into the stomach. The practice began with Australian middle-distance runners in the 1980s and was quickly adopted by some horse trainers. The thinking is this: use an acid buffer to neutralize acid buildup in muscle cells and thus reduce fatigue. Less fatigue, more speed. As with other supplements, scientific studies show no genuine improvement in a horse's performance, and indeed the practice may be harmful to the horse's respiration. The baking soda "milkshake" is banned worldwide and tracks go on testing for it, convinced its use would otherwise proliferate.

Another recent fad has trainers feeding Thoroughbreds so-called muscle builders made from rice oil. Weight lifters have long used this anabolic agent (not a steroid) to increase muscle mass. "Frankly," one equine nutritionist told me, "I think it's manufacturers trying to sell one more thing to horse people, who will try almost anything if they think it will give them an edge over the next trainer."

However misguided, humans have always valued horses. Right from the very beginnings of domestication, horses received better treatment than any other animal. The oldest surviving book on horse care, written by a Mesopotamian trainer named Kikkuli, reveals that in 1360 B.C. horses did indeed get royal treatment: they were bathed, massaged with butter and fed a luxurious diet of grass and grain.

In late nineteenth-century Wales, horses were regarded as the most valuable stock. In his book, *The Role of the Horse in Man's Culture* the anthropologist Harold B. Barclay, now retired from the University of Alberta, notes that horses were served the best food and a wagoner (a servant encharged with driving the farm wagon) who stole grain for his horses was considered an asset to the farm. The wagoner was the aristocrat among farm laborers. The farm owner's eldest son, meanwhile, had the privilege of working with the horses; the second son, the cattle; the third son, the sheep. Significantly, the farmers spoke Welsh to the cows (who got Welsh names) and English to the horses (who got aristocratic English names such as Prince, Captain or Duke). English, of course, was the language of the conquerors.

It was the same in old Hungary. The horse herdsman had a much higher rank than the shepherd.

One national American study in 1991 noted that U.S. courts punish cruelty to animals according to the animal's market value, and in that hierarchy cruelty to horses has long been deemed the greatest sin. Starving a horse generally netted the offender twenty times more days in jail than starving a dog. Respect for the horse is deeply rooted.

But how much did riders and trainers through time actually know about the horse? The truth, it seems, is that some knew a great deal but most had no clue. Xenophon may have been ahead of his time, but horses, sad to say, have seen few of his ilk during the thousands of years they have spent in human company.

I once mentioned to another writer that I was working on a book about horses. His advice was to include a chapter for people like him, who hated horses as a child and who still hate them. This man grew up on the prairie in the 30s and 40s and rode to school on horseback. Clearly, by his sour recollection, the rider had never been trained to ride, nor had his horse been properly schooled. Naturally, neither much liked the other. A little knowledge would have gone a long way.

Artists, meanwhile, clung for centuries to the notion that the horse in a gallop had both forward legs stretched out before him; and the rear, the same. Painters depicted rocking horses, therefore, not real ones. Sculptors of bronze equestrian statues even erred in depicting the horse's walk. Familiarity had bred not so much contempt as a dogged ignorance. We already knew the horse, didn't we?

It appears not. We stand to be surprised. And so it is that *The Nature of Horses*, written in 1997 by Stephen Budiansky, is full of surprises. "It has only been in the last ten years or so," he writes, "that basic science has begun to focus intensively on the horse, and the resulting explosion of research into the evolution, behavior, biomechanics, energenetics, perception, learning, and genetics of horses has yielded remarkable insights into the true nature of the beast."

When I was in Wyoming riding Radish, we played a kind of game. I would occasionally ask that he lengthen his stride in the walk, as opposed to trotting, which he preferred — a matter of saving my seat and his energy. Radish's eagerness to trot, I presumed, had something to do with his general excitability and his desire to be at the front of the pack — which may well have been factors.

But what I learned from Budiansky's book is that new research has shown how a medium trot is actually more energy efficient than a fast walk. Likewise, a medium canter is more efficient than a fast trot. In a very real way, then, Radish — who faced long days, steep mountain trails and taxing desert crossings — was trying, if only I would let him, to do the smart thing. Conserve energy.

Smart is a word not commonly associated with horses. Countless books about horses, even the sympathetic ones, make repeated references to "our dumb companions" and "the noble brutes." In *The Nature of Horses*, Budiansky claims the middle ground, suggesting that horses are neither as smart as, say, Clever Hans was supposed to be, nor as dim-witted as some detractors still claim.

There is an arresting image in his book: a photograph of a robot with what look to be eyes and four legs. Powering it is a great tangle of wires, software and hardware. George Lucas might have wanted to mount a laser on it for use in his *Star Wars* films, though the thing looks more about malarkey than malevolence. The robot is the creation of the Massachusetts Institute of Technology's Leg Lab and illustrates how enormously complicated four-legged locomotion is.

Budiansky argues that we give too little credit to the horse's intelligence considering the amount of information processing involved when a horse canters on uneven ground with a (not necessarily balanced or gifted) rider in the saddle. Cantering may come naturally to the horse, but it is still impressive when you stop to think about it.

A staple of television science programs is the video of the octopus removing the lid from a jar to get at the food inside — proof, we conclude, of the animal's problem-solving ability and therefore of his intelligence. Carnivores (mostly) ourselves, we prize carnivore logic. Octopi and horses, it turns out, do equally well on tests involving a choice between two visual patterns to get at a food reward behind one of them. The horse's reputation for memory, it seems, is richly deserved. A horse, for example, after learning twenty pairs of signs (the food reward is always behind the cross, never the circle; the L, never the R) remembered them all on retesting, and even twelve months later demonstrated almost perfect retention.

Memory is critical for a grazing animal whose survival hinges on recalling where that watering hole in the high desert was or where the grass in one sheltered valley stayed green long into winter. Horses are not uniformly adept at problem solving (a simple maze test in which a left turn led to a reward and a right turned to a dead end still confounded 20 percent of horses after five trials), but they don't have to be. Carnivores must find *and* catch their prey, which will try to elude them in myriad ways; herbivores must simply locate food. Different problem, different mind-set. As Budiansky says, "Mice move and hide, grass doesn't."

Unlike the carnivore, the horse's primary tasks are to reproduce and, when danger presents, to flee. But even fleeing, as the MIT Leg Lab knows all too well, is tricky business.

When Columbus brought horses to the New World in 1494 on his second voyage, the twenty-four stallions and ten mares on board his fleet were actually returning to the land of their ancestors. The horse first evolved in North America: the forerunner was a little doglike creature that anthropologists now call *Hyracotherium*, or *Eohippus*, though I like what they used to call it in the dinosaur books I devoured as a child — the "dawn horse." Over time, as the climate changed and got drier, the forests gave way to grassland in the center of the continent. And although Incitatus and Morzillo got chicken, grass, of course, is what horses love to graze on.

As the savannah grew more lush, the little dawn horse got bigger. The prehistoric horse could no longer hide as he had from his enemies in the forest; he became a creature of flight, and his legs grew longer. His several toes evolved into one toe — a hoof. Over time the dawn horse's teeth changed to handle new vegetation: not leaves and fruit but grass, and lots of it. He would need powerful front teeth to clip the grass, and molars to chew it. To make room for these teeth, the horse's face grew longer, and eyes set on the side of his head gave him 360-degree vision to spot his enemies.

The result was *Equus caballus*, the horse we know today. *Equus* thrived on the Great Plains, then a seemingly endless pasture. Using land bridges that no longer exist, some horses crossed over to what is now Europe and Asia. But about fifteen thousand years ago all the horses in North America — along with camels, saber-toothed cats, mastodons and some

3.2   Cave painting in France: the horse was a source of both food and
inspiration to our ancestors.

other creatures — disappeared. No one is sure why. Was it climate change or massive flooding? Did primitive humans hunt
horses to extinction? Or did disease do them in? For eons, our
continent was horseless.

(Or was it? A controversial "lingering herds theory," proposed by some anthropologists, argues that pockets of horses
survived the Pleistocene. These academics point to the fact
that the Dakota-Lakota people between the Mississippi and
the Rockies were, even in the early 1700s, "extremely bold
and daring riders," according to French explorers' accounts.
It seems hardly possible, given the limited numbers of Spanish
horses then on the continent. But until the bones of post-
Pleistocene/pre-Columbian horses can be found and dated,
the lingering herds theory will remain just that.)

Meanwhile, on the plains of Russia, primitive horses continued to survive and spread into present day Europe. But
here, too, they were a source of food long before they were
ever domesticated or ridden. Near the French village of Solutré,

scientists have found a three-foot-deep layer of horse bones stretching over several acres. The remains of some one hundred thousand horses lie here. Their skins and dung and flesh would have offered warmth and firelight and food.

Equus might have disappeared in Asia and Europe, as in North America, had someone with a bold imagination not demonstrated *other* uses for the horse. Perhaps early humans gave the horse a role to play in a rite of passage or some spiritual quest. Domestication of the horse would then have led to practical uses in transportation and war. Humans have caused the demise of many species, but by our bits and bridles and harnesses we may have actually saved one from extinction.

The long evolutionary road from *Hyracotherium* to *Equus*, or even Columbus's oceanic path from Spain to Hispaniola (Haiti), was, however, no simple trajectory. If it is true that there is nothing new under the sun, it is also true that nothing is as it seems.

Let's start with Columbus. Growing up in the 1950s, I was taught in geography class that flat-earth fearmongers wearing the scarlet robes of a cardinal tried to dissuade Spanish royalty from backing Columbus on his journey. They worried, as sailors did, about going over the edge.

In an essay entitled "The Late Birth of a Flat Earth" (collected in *Dinosaur in a Haystack*), the evolutionary biologist and paleontologist Stephen Jay Gould tries to fathom why he was taught in the 1950s this same myth: that for a thousand years in the so-called Dark or Middle Ages, scholars believed the earth was flat.

The tale painted Columbus as brave seeker; the clerics, as narrow buffoons warming up for the Inquisition. Science versus Religion. Light versus Dark.

That pungent parable lodged in my brain. But little of the story is true. Aristotle and the Greeks — several hundred years before the birth of Christ — knew the earth was round, as did both scholars and clerics long before and during Columbus's time. When Columbus approached Ferdinand and Isabella of

Spain, looking for patronage, the clerics did indeed question him. They grilled him on his figures, convinced he had underestimated the circumference of the earth, and they were absolutely right.

The flat-earth myth, Gould argues, took hold in the late 1800s in Britain when certain intellectuals needed "whipping boys and legends" to advance their theory of history as the victory march of enlightened science over grim religion. The story fell into the history books like an errant mosquito trapped between pages when a heavy book is snapped shut.

As for the breeding stock Columbus brought to the New World, the horses were less fine than imagined. It seems that before boarding ship the officers sold the purebred steeds they were supposed to bring in order to buy wine for the journey. What little money remained was used to buy some cheap and sorry horses. Only six survived the awful crossing. Crammed into small ships, the horses were hoisted with straps under their bellies and suspended from the ceiling of the lower decks for some forty days and forty nights. Typically, a third of them perished on these journeys.

(Later, as other ships — with horses on board — sailed to the New World, their sails sometimes went slack in the often wind-free zones at Cancer thirty degrees north latitude and Capricorn thirty degrees south latitude. When water supplies ran out and the horses went mad with thirst, sailors shot them and tossed their bodies overboard to the sharks. The regions are still called the horse latitudes.)

Even after death, the Italian boy who had married a Portuguese and sailed for Spain went on journeying: Columbus's bones were disinterred many times over the centuries and lie now, it is claimed, in the great cathedral of Seville. I have stood before the high, elaborate tomb and can report that he lies flat, as flat as the earth was once purported to be.

As for the neat evolutionary line from *Eohippus* to *Equus*, here, too, Gould reminds us that Darwinian evolution is not *just* a slow, steady going to the light, with each species improving and adapting through time. Gould and many of his colleagues believe that if the evolution of horses, humans or dinosaurs is seen as a tree or bush, "Trends surely exist in abundance, and they do form the stuff of conventional good stories. Brain size does increase in the human bush; and toes do get fewer, and bodies bigger, as we move up the bush of horses. But the vast majority of bushes display no persistent trends through time."

Gould argues for something he calls "stasis": the notion that evolutionary change occurs comparatively quickly and in small, isolated populations, but that for glacially long periods, no change at all is the norm. The history of *Equus* would seem to back his contention: until twenty million years ago, when the horse branched out into variously sized versions, the body size of ancestral horses had remained a virtual constant for thirty million years.

It would also be wrong to view certain branches of the equine tree as evolutionary failures, or to see *Equus* as evolutionary perfection. At one point in the vast history of the horse, thirteen species existed simultaneously, some big, some small and all, in their own way, successful. We do like a good story — witness the revisionism around Columbus — preferably a tidy, romantic one. The evolutionary truth is infinitely more complex, and more interesting, than that.

In the meantime, we should respect both horses and dinosaurs, who have proven their mettle in that harshest of tests, the test of time. When a Yukon government archeologist named Ruth Gotthardt was asked in 1993 to investigate the carcass of a horse found frozen in the ice — it turned out to be a twenty-six-thousand-year-old horse, and remarkably like contemporary horses — she initially had grave doubts about the specimen's age. What greeted her nose when she descended into the trench that day was "the unmistakable smell of horse droppings." Gotthardt was astonished that "something from the Ice Age could keep that smell of horse." Call it an enduring smell, and a tidy metaphor for longevity.

——

In considering the long and stalwart history of horses, and the fact that there are now some sixty million horses on the planet, we must also give a nod to lady luck.

Budiansky, in *The Nature of Horses*, comes to grips with a most intriguing question. Why is it that humans managed to domesticate horses (along with cows, sheep, dogs, cats, pigs and goats) and not other creatures? The ancient Egyptians attempted the same thing with antelope, gazelles, hyenas and ibex. No luck. American Indians apparently kept pet raccoons, bears, even moose, while aborigines in Australia did the same with wallabies and kangaroos. No luck there, either.

As chance or fate would have it, the horse actually seemed designed to accommodate some sort of partnership. Small wonder the ancients saw the horse as a gift of the gods. Capable of surviving on meager fodder that would soon starve a cow, the horse also possessed a gap — called a "diastema" — between the front teeth and the back molars: a perfect place for the bit, in turn attached to bridle and reins. The horse, also happily for humankind, was a social creature whose complex and silent language, with all its signals for aggression and submission, enabled the animal to understand similar human signals. In other words, you could train *Equus*. You cannot always do that with other animals, even equine cousins. Trainers trying to work with young zebras in harness typically report calamitous results.

Stories of intractability abound about Przewalski's horse, a wild cousin to the horse and once native to China, Russia and Mongolia. This breed is now presumed extinct in the wild after the last one was spotted a few decades ago. The horse owes his name to a Polish-born officer of the Imperial Russian Army, Colonel Nikolai Przewalski, who shot one on the China-Mongolia border more than a hundred years ago and shipped the skull back to a zoological museum in St. Petersburg. Some twelve hundred individuals — descendants of eleven caught at the turn of the century — still exist in zoos and captive breeding centers around the world. Today, about forty "P-horses," as they are sometimes called, roam a one hundred-and-fifty-thousand-acre reserve in Mongolia, where Przewalski's horse remains the national symbol.

3.3   Przewalski's horse: a member of the equine line that dates back 55 million years.

Only twelve to fourteen hands high, with a dorsal stripe and spiky black mane, these horses have a reputation for intensity, aggression and wariness; they look strikingly similar to the horses drawn by our Stone Age ancestors on cave walls fifteen thousand years ago.

In free-range pastures, Przewalski's horses will form a circle — as musk ox and zebras and wild ponies do — against wild dogs or other predators, and both mares and stallions will menace human or animal interlopers. In one account, a small Przewalski's stallion rather savagely dispatched a bigger domestic stallion who sadly underestimated his opponent. With one exception (a horse at the San Diego Zoo), none of these primitive horses, as far as I know, has ever been tamed or ridden. Their reputation for aggression, though, may be inflated: the zoologist Lee Boyd, who has studied them in Mongolia, told me they are no more aggressive than the mustangs of the American plains.

Finally, and this may be the most significant element of all in why the horse-human bond developed, recent evidence from archeology and animal behavior studies suggest that the partnership was as much the horse's idea as the human's. Like mice and rats, raccoons and cockroaches, horses may have learned that while hanging around human settlements came at a cost to a few unlucky individuals (who became horse stew), the overall advantages for the herd far outweighed the disadvantages. Predators were less inclined to prey on them so close to villages, and the crops the horses raided offered better-than-usual fare.

Domestication was a natural consequence, but only for those horses, as Budiansky puts it, who were "more curious, less territorial, less aggressive, more dependent, [and] better able to deflect human aggression through submission."

The best archeological evidence — from the wearing on horse teeth due to crib biting — suggests that the habit of corralling or tethering horses may go back thirty thousand years. Evidence for the first rider, again from the latest dental remains, locates that moment on the Russian steppes some six thousand years ago. But Harold B. Barclay, the anthropologist, holds out an intriguing possibility: perhaps early riders had no need of bits but rode skin to skin as the Plains Indians more or less did.

Barclay cites a report from 1885 of a French cavalry officer who mounted a spirited horse without saddle or bridle, galloped across a field, jumped over tent ropes, trenches and other barriers, turned and stopped the animal at will. The officer used a strap around the horse's neck as brakes, and to turn he slapped the horse's neck with his palm.

On the one hand, that officer had the benefit of centuries of horsemanship. On the other hand, much of that horsemanship was dubious, and his gift with horses may have been beyond the ken of most of his and our contemporaries. I think of Monty Roberts, a trainer who learned the essentials of his gentling craft as a thirteen-year-old boy simply by watching wild horses in the high desert of Nevada in 1948. No one *told* him you could gentle a horse; indeed, conventional wisdom advised aggression. He went the other way, and he wonders whether others ever did the same. "Horses are such good teachers," he says. "There *had* to be people before me."

We may never know whether individuals or whole Ice Age tribes could ride as that nineteenth-century officer could, *sans* bit or bridle or saddle. That being so, we may never know precisely when that first rider felt the bracing wind that comes only on the back of a fast horse.

Some things we like to think we know. By 1286 B.C., the Hittites and Egyptians were clashing with the tanks of their day — horse-drawn chariots, many thousands of them. In 950 B.C., King Solomon would boast of twelve thousand horses in his stable. (Solomon reorganized the army in Israel and Judea and may have built a huge chariot force. But the number twelve in the Bible, scholars say, symbolizes completeness, so that twelve thousand may constitute a plucked number. Proof again that things are seldom what they seem.)

Next came warriors on horseback, though it must have been amusing to watch them mount. Did they step up on a block? Use their spears as pole vaults? Or did the horse kneel like a camel?

Between 500 B.C. and A.D. 400, Hun tribesmen of the Russian steppes invented both the saddle and the stirrup — opening up new possibilities for warfare. The stirrup, according to some observers, was the most significant military development in the five-thousand-year span between the taming of the horse and the inventing of gunpowder. It is easy, then, to see the history of the horse as the history of war.

The war horse was often sacrificed in great numbers, especially as weapons of war became more deadly. But a special affection between certain soldiers and certain horses is apparent all through the ages, and though early humans made a meal of the horse — and horsemeat is still prized today in France and Belgium, for example — some societies simply will not consume horseflesh.

Buddha, like one pope, prohibited its consumption. Muhammad never banned it but never ate it, either. But in Victorian England, out of concern about the protein-deficient diet of poor people, The Society for the Propagation of Horseflesh as an Article of Food was formed in 1868.

A much publicized dinner at the august Langham Hotel in London offered a nine-course meal that included horse soup; fillet of sole in horse oil and fillet of roast Pegasus; turkey with horse chestnuts; sirloin of horse stuffed with Centaur; braised rump of horse; tongue of Trojan horse and jellied horses' hooves in maraschino.

The push to rediscover horsemeat occurred at a time when the price of beef was soaring and epidemics plagued the cattle industry — shades of mad cow disease. Victorian England required millions of horses for the transportation of people and goods, so it did not lack for horseflesh; still, horsemeat never caught on. The Christian taboo was too entrenched, feeling for the horse too strong.

———

In the New World, an ocean away, the horse would enjoy a rebirth in the clearing of the land, the drama of the buffalo hunt and the centuries of fighting between cowboy and Indian. The great migrations of families across the continent in covered wagons and Red River carts, and of cattle from Texas to the north and west, could not have occurred without the horse. This was a new chapter in horse and human history.

With dizzying speed, the horse that Columbus brought spread north and west and plains riders developed exquisite skills. George Catlin, the nineteenth-century traveler who mingled with the Comanche in 1834, was at first confounded by the war games of young Comanche boys. Here is what he saw: at a full gallop, the boy would streak past an imaginary enemy, dropping down on the far side of his animal so only one heel and perhaps the top of his head were visible. From that position (all the while toting lance, shield, bow and quiver), he would let loose arrows either over the pony's back or under his neck. Neither pony nor rider came to grief, and the arrows landed on target.

3.4   Plains warriors, in short order, developed into the finest light cavalry in the world.

3.5   Hunting a tiger in India: humankind would find many uses for the horse.

Curious to know how this was done, Catlin offered one boy some tobacco to get a closer look. What enabled the circuslike maneuver — in part, anyway — was a hair halter braided into the mane at the withers. That loop cradled the boy's elbow and took his forward weight. But the balance demanded the deft use of the heel — this required years of practice.

If, during a skirmish, the warrior's horse was wounded or killed, the rider invariably landed on his feet and used his shield as protection. He then waited for help. It usually came, since Comanches felt a deep compulsion to rescue wounded or dead comrades: the wounded because it was the honorable thing; the dead because a scalped human had no place in the hereafter.

Such rescues demanded extraordinary horsemanship. Young boys would train for the maneuver by first learning to ride hard toward a small, light object on the ground, lean down and scoop it up. The objects got heavier, bulkier, and then two riders galloping side by side would haul up a boy playing the role of wounded comrade, his hands in the air to help the lift, and one rider would lay the boy across the horse and continue on. Finally, the boys would practice picking up a "dead" comrade — who could be on his belly, his back or crumpled in a heap — while riding full out.

Stanley Noyes, the author of *Los Comanches: The Horse People*, noted that "This extraordinary maneuver required perfect timing, great agility and strength, and expert riding. If performed today during a rodeo, it would probably leave even our hypothetical American cowboy shaking his head. Yet every Comanche warrior was supposed to be able, if the need rose, to effect such a rescue by himself, and this at the risk of his own life from enemy arrows or gunfire."

Although the Comanche felt no great affection for horses in general, each warrior had his favorite war or buffalo horse, who would be extricated from the herd and staked next to his lodge at night. Like knights and Arab warriors of old, he rode a common horse *to* battle but his prized war horse *into* battle. If ever that horse (who often doubled as a buffalo horse) was sold, he would fetch an object of veneration, such as a white buffalo hide. In other words, that horse was almost priceless. "Some men," wrote Noyes, "were said to love such animals more than their wives or children."

The story of the horse in Canada is the story of humble beginnings and horse fever taking hold — among both indigenous peoples and newcomers. The first colonial horses came in 1665 from the royal stables of Louis XIV. Eighteen mares and two stallions would form the basis of the hardy black horses later called the Canadian breed. Those who survived Quebec winters became legendary as stalwart all-purpose horses and easy keepers. So attached to these "little horses of iron" did French colonists become that when the English were about to attack in 1757 and food supplies dwindled, none would slaughter their Canadian horses.

Early census figures show that in 1681 there were only ninety-four known horses in the entire country. Then the numbers soared: 156 in 1685; 580 in 1695; 5,270 in 1720; 13,488 in 1765. The government grew alarmed at the rapid increase and restricted each farmer to two horses and a colt lest the horses interfere with the raising of cattle. In 1710, the governor complained that there were so many horses in Canada that "young men were losing the art of walking, with or without snowshoes." His solution, a time-honored one, was to kill some of the horses. They could, he suggested, be salted and sold to the savages "*en guise de boeuf.*"

On the plains, it was "the savages" who provided Manitoba settlers with mustang horse power, and the pioneers were glad of it. Still, they longed for bigger horses to pull their plows. Enter the legendary Fireaway — a sixteen-hand red roan stallion. He was a Norfolk Trotter who endured not only the long sail across the Atlantic but also the journey from York Factory on Hudson Bay to Fort Garry (Winnipeg) — in a canoe. Somehow the horse learned to stand and balance himself, even in rapids, and obligingly exited and entered the canoe on portages.

No one was disappointed when Fireaway arrived in his new home. He would change the stock of western horses for decades. The Wonder Horse of Red River, they called him. Fireaway the Marvelous. Handsome, fast and apparently a target for horse thieves, he won the admiration of everyone. A Hudson's Bay Company governor wrote: "He is looked upon as one of the wonders of the world by the natives, many of whom have travelled great distances with no other object than to see him."

As recently as the 1920s, to be a farmer in Canada was still to be a horseman. Allegiance to a particular breed of plow horse, like allegiance to a particular political party, was passed on within the family. British, Scottish and Irish farmers tended to favor Clydesdales; settlers from continental Europe and America preferred the bigger Percherons and Belgians. In Saskatchewan, then home to one million horses, rural churches would divide on a Sunday morning, the Clydesdale backers on one side, the Belgian and Percheron faction on the other.

Among Indian cultures in what is now Canada the horse almost instantly became currency. Among the Blackfoot, Blood and Piegan, one to twenty horses was the price of a wife. Adultery by a wife sometimes led to the cuckolded husband's being paid a horse (and the woman's losing her nose, which was cut to the bone).

Among the Blackfoot, a warrior's horses were slain so he could ride them to the spirit world, as was the custom in many parts of the Old World. When one great chief among the Blackfoot died, 150 horses were slaughtered. The Assiniboine, meanwhile, gave a dead warrior's horse his freedom. Horses even entered the Assiniboines' dreams. One early anthropologist observed that his Assiniboine interpreter did not dream of things to eat. He preferred, instead, to dream of horses.

CHAPTER 4

THE GENTLE ART OF THE HORSE WHISPERER

There was never a king like Solomon
Not since the world began
Yet Solomon talked to a butterfly
As a man would talk to a man
RUDYARD KIPLING

KING SOLOMON, as far as we know, could not literally talk to the animals, but his legend speaks of our ancient and heartfelt desire to close the communication gap between us and them. Fiction, poetry and myth teem with characters — King Solomon and Dr. Dolittle among them — who *could* talk to the animals and animals who could talk to one another.

Forty-six years ago the noted animal behaviorist Konrad Lorenz wrote *King Solomon's Ring*, the title a reference to the magic ring that supposedly allowed Solomon to converse with the creatures. The legend of the ring may stem from a misreading of the Bible: Solomon "spake also of beasts, and of fowl, and of creeping things, and of fishes." Spoke *of*, not *to*. (Kipling's time in India, with its rich Muslim folklore and the *Arabian Nights*, where Solomon is a character, more likely inspired the legend.)

Coyly, Lorenz professes to believe the legend, ring or no ring. If he, Lorenz, can talk to the animals, why couldn't Solomon? The title page of *King Solomon's Ring* includes a sketch of a man with his hands in his pockets, bending low to chat up a duck.

"Without supernatural assistance," writes Lorenz, "our fellow creatures can tell us the most beautiful stories, and that means *true* stories, because the truth about nature is always far more beautiful even than what our great poets sing of it, and they are the only real magicians that exist."

Animals have long told stories, but few humans have felt inclined or taken the time to listen. Consider, for example, the manner in which we have schooled the horse for thousands of years. "Breaking" a horse constitutes a pivotal moment in the life of that horse. First bridle, first saddle, first rider, like all first impressions, surely shape the horse.

If you're looking for an image of how we have traditionally undertaken that schooling, look no further than film. Consider *Monte Walsh*, a cowboy film released in 1970 and packed with clichés, not least the one about broncs and broncbusters.

Lee Marvin is Monte Walsh, a noble drifter who will not settle down even as the cowboy era fades and, with it, ranch work for men like him. What money he has he gives to the gold-hearted hooker, played by Jeanne Moreau. "It ain't like I ain't got a horse," he says, implying that he's not *that* poor. Monte's pal, the hollow-cheeked Jack Palance, hangs up his chaps to marry the hooker and run the hardware store in the town of Harmony.

When Shorty, cowpoke gone sour, kills Palance's character in a robbery, Monte stalks the former to get revenge, but

4.1   Traditional bronc-busters went from ranch to ranch, charging $5 a head to make a wild horse manageable.

not before breaking a gray stallion — in an extraordinary scene. The corral, the inside of a store and numerous storefronts splinter before the bronc is conquered. "I rode down the gray," Monte tells the dying Shorty, rubbing salt in his considerable wound, for Shorty had earlier found the grey too hot to handle. The phrase, "to ride *down*" a horse, vividly expresses how green horses were broken in the Old West, and even today in the New West. The Wyoming license plate features a blue silhouette of a cowboy on a bucking bronc.

Still, much has changed. The 1990s, especially, have seen the ascendance of horse gentlers, historically called "horse whisperers," and many of them, ironically, from the same territory that spawned the likes of Monte Walsh. As one young western horse trainer put it, "I had to give myself a macho-ectomy."

Monty Roberts, the author of *The Man Who Listens to Horses*, knows Hollywood film sets almost as well as Lee

Marvin. He spent his youth working on them as wrangler and double — for Elizabeth Taylor, Charlton Heston, Mickey Rooney and others in a hundred or so films such as *National Velvet*, *East of Eden* and *My Friend Flicka*.

But Roberts would take no pleasure in watching the breaking of the gray. He is more of the Solomon school. So are his compatriots Tom Dorrance, Ray Hunt, Buck Brannaman, Richard Shrake, John Lyons, and Linda Tellington-Jones. "Broncobusters Try New Tack: Tenderness," went a *New York Times* headline in 1993.

Things are changing, and not just on ranches. Monty Roberts is also a world-class trainer of Thoroughbreds (his horse Alleged won the prestigious Prix de l'Arc de Triomphe two years running) who thinks whips should be banned from the racetrack. European show jumpers who use their crops on horses when they balk at fences — once deemed fair punishment — are now met with a chorus of disapproving whistles. And if the gentlers and whisperers hold sway, spurs and crops and rough bits may one day find a place in horse museums, not in stables and barns. Alas, we are a long way from that, as a glance back at centuries of horse breaking will attest.

Yet here is the curious thing. You can draw a line from modern-day gentlers right back to the ancient Greeks. Through time, infinitely more trainers have been inclined to the Monte Walsh method than to the Monty Roberts method, but you can find gentlers if you look. Those who love horses will anguish to know that the essential wisdom of the horse whisperer is more than two thousand years old. "A good trainer can hear a horse speak to him," says Monty Roberts. "A great trainer can hear him whisper." (In folklore the trainer, not the horse, did the whispering. It was believed that a magic word was known only to a select blacksmith-led brotherhood. That word, when whispered in the ear of even a wild stallion, would tame him.)

The skills required to hear the whisper were there all the

time and largely ignored. Why? Why did gentling horses fall in and out of fashion? And why have we rediscovered it now?

Three hundred years before the birth of Christ there lived the Greek cavalry officer Xenophon (meaning, "a person who speaks strangely"). A pupil and friend of Socrates, he wrote a tiny classic called *The Art of Horsemanship*.

I first encountered his ideas ten years ago. At the age of thirty-eight, I took up riding for a very practical reason: I was working with the show jumper Ian Millar on his memoirs and wanted firsthand experience in the saddle. Knowing this, a neighbor passed me her old hardback copy, its cover a faded peach brown, of *Horsemastership: Methods of Training the Horse and the Rider*, written by the American author Margaret Cabell Self in 1952. In her preface, she sang Xenophon's praises and stressed that almost everything he recommended is as pertinent today as it was in his time.

"Nowhere," wrote Self, "did he recommend violence or the use of fear to subdue the animal. One could wish that those who use brutality in breaking and schooling a horse — the method of throwing a horse, tying him up, wearing him out, or flogging him over jumps — that these products of modern civilization who take pride in the strength which has enabled them to 'break' and subdue their mounts could have the understanding of the ancients."

As I struggled to learn the grammar of his sport, Ian urged me to read the modern-day bible of classic riding, *Hunter Seat Equitation*. The title struck me as bizarre. How could those three words have anything to do with one another? (A hyphen, as in Hunter-Seat Equitation, might have helped.) That book, written in 1971 by the American show jumper George Morris, features a note of thanks to Margaret Cabell Self — "for putting me on a horse." And like Self in her book, Morris in his rails against the use of spurs and crops by "butchers" who call themselves riders. Sometimes, it seems, students actually listen to their teachers.

Regrettably we stopped listening to Xenophon. Be firm

4.2   Rodeo girl and horse, 1912: horse-human history is full of lessons
learned, then forgotten.

but not harsh, Xenophon urged, and *never* lose your temper
while dealing with a horse. "A fit of passion," he wrote, "is a
thing that has no foresight in it, and so we often have to rue
the day when we gave way to it. Consequently, when your
horse shies at an object and is unwilling to go up to it, he

should be shown that there is nothing fearful in it, least of all to a courageous horse like him; but if this fails, touch the object yourself that seems so dreadful to him, and lead him up to it with gentleness."

The irony is comic. Xenophon trained horses for war, that most violent of mass enterprises, but he cautioned riders that aggression (at least with horses) does not pay: "Reward him with kindness when he has done what you wish and admonish him when he disobeys." On the other hand, "riders who force their horses by the use of the whip only increase their fear, for they then associate the pain with the thing that frightens them."

If Xenophon's horsemanship marked an island of enlightenment, the dark ages that preceded and followed him were indeed dark and perilously long for the horse. Greek chariot drivers of the sixth century B.C. deployed bits with spiked cheek pieces that offered the driver control, all right, but at the expense of the soft tissue around the horse's mouth. The habit of not castrating stallions and the absence of sophisticated training methods may have made such bits necessary.

The ancient Egyptians, meanwhile, would control spirited horses by using an extremely low noseband. Unlike today's drop noseband (which sits three inches above the nostrils), the severe ancient version was often attached directly to reins on either side. This arrangement not only put terrific pressure on sensitive tissue right at the horse's nostrils but also interfered with breathing (the horse cannot breathe through his mouth). The rather savage solution was to slit the horse's nostrils. Bas-reliefs from 1450 B.C. testify to the practice, which resurfaced in Europe between the fifteenth and seventeenth centuries, when similarly low nosebands were also used.

The decline of the Greek civilization and ascendancy of the Roman would also spell hard times for horses. Roman trainers, bronc-busters in togas, would force the horse to the ground, sit on his head and tie his legs together. Xenophon's book seems not to have made it to the imperial capital.

Perhaps the notion of gentling war horses struck the Romans as unseemly, even unmanly. The same harsh discipline imposed on legionnaires and conquered tribes was inflicted on horses. For centuries, no one challenged that philosophy.

In tenth-century Iceland, sagas describe how stallions were made to fight each other, as pit bulls and cocks are today. Even thirteenth-century Mongols under Genghis Khan, who professed their love of horses and felt real affection for them, nevertheless broke them harshly — that is to say, traditionally — and rode some of them to death on cross-continental raids.

No one gave much thought to setting down a *system* for training horses until along came Federico Grisone, the six-teenth-century horsemaster from Naples. His book, *Gli Ordini di Cavalcare*, inspired riding schools all over Europe. An old woodcut shows him in feathered hat and spurs, holding a sharp pointed stick against a horse's chest.

Gentle, Grisone was not. "In breaking young horses," he wrote, "put them into a circular pit; be very severe with those that are sensitive and of high courage: beat them between the ears with a stick."

The horse afraid of crossing streams, Grisone advised, should have his head forced underwater. The bits he devised were brutal. Grisone took seriously his role as subjugator of the horse and his words attest to a cold confidence in that approach: "The 'breaking down' or 'taming' is not without the most desperate trial on the part of the horse, which rears and plunges in every possible way to effect its escape, until its power is exhausted and it becomes covered with foam; and at last it yields to the power of man and becomes his willing slave for the rest of his life."

This makes for distressing reading, and yet I take solace in the knowledge that sometimes young trainers do *not* listen to their elders; they listen, instead, to the horse. Monty Roberts's father, for example, responded viciously when his son re-nounced the old way (in essence, the Grisone way). The father beat the son so severely with a stall chain that Monty had to be treated in hospital. Monty, though, persisted in gentling horses as a young teenager in the 1940s.

Similarly, Grisone's teachings in the sixteenth century were foisted on, and rejected by, Antoine de Pluvinel, who became Master of Horse to the king of France, Louis XIII. Like Xenophon, he urged that the trainer calmly inspire confidence in the horse and that each horse be treated as an individual.

Contemporary Englishmen thought otherwise. Thomas Blundeville recommended that riders be equipped to deal with recalcitrant horses: an iron bar set with prickles would be suspended from the horse's tail and passed between the horse's legs by a cord. The rider could thus draw the cord up and mete out punishment whenever he saw fit. If this failed, Blundeville advised, "let a footman stand behind you with a shrewd cat tied at one end of a long pole with her belly upwards, so as she may have mouth and claws at liberty. And when your horse may stay or go backwards, let him thrust the cat between his legs so as she may scratch and bite him, sometimes by the thighs, sometimes by the rump and often times by the stones."

The Renaissance (the revival of art and literature between the fourteenth and sixteenth centuries), meanwhile, had at least elevated riding into an art form. An educated person was expected to know something of music, art, literature — and equestrian matters. Methods used to instruct in *haute école* horsemanship may not always have been gentle, but there were glimmers of light.

By the eighteenth century, the duke of Newcastle was setting up riding academies in Paris and Brussels and his methods were a far cry from Grisone's and Blundeville's. "A boy," he wrote, "is a long time before he knows his alphabet, longer before he has learned to spell, and perhaps several years before he can read distinctly; and yet there are some people who, as soon as they get on a young horse, entirely undressed and untaught, fancy that by beating and spurring they will make him a dressed horse in one morning only. I would fain ask such stupid people whether by beating a boy they would teach him to read without first showing him the alphabet? Sure, they would beat him to death, before they would make him read."

Meanwhile, over in the New World, footloose horses — who had escaped the Spanish or the indigenous peoples who stole them — thrived on the rich buffalo grass of the prairie. Wild horses spread north and west and those who captured them became lords of the plains in relatively short order.

Plains Indians tended to be firm as horse breakers but also slow and easy (unlike cowboys, who seemed to be in a terrible hurry.) Even the moment of capture — in winter and early spring, when mustangs were weak, or at watering holes after they had gorged themselves — showed patience and forethought. Another tactic was to exploit the herding instinct of horses: their tendency to move in a wide circle when flushed. The tamer's task was to keep them moving in that circle until they were exhausted.

Like modern horse whisperers, the Indian horse breaker believed touch was important. He would walk toward the roped horse, talking all the while and eventually touching the horse's nose and head.

He would also breathe into the horse's nostrils, an ancient practice rooted in the knowledge that wild creatures smell first, trust second. (Mustangers and horse trainers soon learned not to change their clothes, even their underwear, while working with horses over the course of days — to avoid starting over. The horse's keen sense of smell, which can find water in the desert, would catalog and tolerate the familiar while rejecting the new.)

A leather halter was then applied in such a way that even a slight pull back by the horse exerted terrific pressure on nerves at the neck and nose. The horse soon got the message and then the warrior could begin schooling.

Uttering deep grunts (*hoh-hoh-hoh*) or calming sighs (*shuh-shuh-shuh*), he ran his hands over every inch of the horse's body. If the horse balked at being touched around the flanks, a tug on the halter would serve as a reminder to stand still. The aim was to convince the horse that no harm would be forthcoming.

Following all this, actual riding posed no real difficulty. It strikes me that while the training of a horse can reach immensely complicated dimensions, in the beginning, at least, it is a question of simple manners. The Indian trainer made a point of introducing himself to the horse he would ride.

Chief Buffalo Child Long Lance, a Blackfoot, had a brilliant phrase to capture the importance of marking equine courtesies: in *Long Lance*, published in 1928, he advised that wild horses "must first be treated gruffly — but not harshly — and then when he is on a *touching acquaintance* [the italics mine] with man, kindness is the quickest way to win his affections."

Mounting was done slowly and by degrees. The rider approached the near side of the horse and pressed down on the horse's back, first lightly, then harder. With elbows laid across the horse's back, the rider lifted himself off the ground a few inches until the horse bore the rider's weight. Finally, the rider slipped his leg over the horse's back. Astonishingly, few horses bucked.

And a good thing, too. The rider had neither saddle nor stirrup to help keep his seat. The saddle was no more than a pelt cinched under the belly. The bridle was a leather thong looped around the horse's lower jaw. The rider's primary aid was the knees.

Even young Indian boys, the Sioux most notably, broke mustangs, though they wisely cushioned their falls by riding into a lake or river. Using water to soften the wildness is a clever idea, one perhaps passed on. During the First World War a reclamation camp was established in England to handle "untamable" army horses otherwise destined for the slaughterhouse. The demon horses were entrusted to the daughters of wealthy country gentlemen, horse-wise young women who had lived around horses from childhood. The ladies did as

Sioux boys had done: they rode into deep water. Without fail, the unruly horses soon turned quiet and tame.

But while Indians broke horses their way, and cowboys theirs, John Solomon Rarey developed his own unique approach to horses. Much as Monty Roberts would take the horse world by storm more than a century later, Rarey made a name for himself in the 1850s. He was a gentler of sorts, neither bronc-buster nor whisperer but something in between. Rarey wrote of "conquering" the horse through the odd crack of the whip, but he also stressed the importance of a kind voice and touch and insisted, like Xenophon, that fear and anger had no place in the training of horses.

Rarey grew up on a farm in Ohio, where he broke horses the old way — and almost every bone in his body. "His pluck," one observer noted, "was greater than his science." But maybe not, for he had the good sense to seek alternatives by asking around — of cowboys, other horsemen, circus trainers — and to read widely. Eventually, he devised a system of training horses that relied ultimately on gentleness, fearlessness and simple devices (such as soft leather hobbles).

A British cavalryman who witnessed a Rarey demonstration was so impressed he wrote letters of introduction to prominent families in Britain. A plan took shape: Rarey would teach his techniques but only after five hundred clients had anted up more than ten pounds each — a fair sum in 1858.

While waiting for his British equestrian class to top up, Rarey crossed the Channel to tackle a French horse, called Stafford, who had refused to let anyone groom him for a year. Trainers in Paris had tried blindfolding him, muzzling him and hobbling him, but he still attacked anyone who came close. His despairing owners were about to have him destroyed and had nothing to lose by letting Rarey have a go.

A large crowd of horse fanciers gathered to witness the inevitable — Rarey's humiliation. But that is not what the *Paris Illustrated Journal* would report: "After being alone with Stafford for an hour and a half, Mr. Rarey rode on him into

the Riding School, guiding him with a common snaffle-bridle.
The appearance of the horse was completely altered: he was
calm and docile. His docility did not seem to be produced by
fear or constraint, but the result of perfect confidence. The
astonishment of the spectators was increased when Mr. Rarey
unbridled him, and guided the late savage animal with a mere
motion of his hands or indication with his leg, as easily as a
trained circus-horse. Then, dashing into a gallop, he stopped
him short with a single word."

The challenge awaiting Rarey in England was a stallion
called Cruiser, a steeplechaser handled by a groom with a blud-
geon and reputed to be the most vicious horse in the country.
The horse had not been ridden in three years and wore an
eight-pound iron muzzle. His owner conceded it was "as much
as a man's life was worth to attend to him." But attend to him
Rarey did, with infinite patience, perseverance and savvy.

First he rode the horse, though tacking and mounting took
three hours. Then he put the horse behind a cart and walked
him forty miles to tire him. Cruiser was fitted with a gag bit,
straps and hobbles that would render him helpless but would
not cut him. Then Rarey gentled the horse, talking softly to
him and touching him all over. As one observer put it, he
sought to tame him "limb by limb, and inch by inch."

If Cruiser displayed any sign of hostility, Rarey would lift
the horse's still-helpless head and shake it, as a father would
scold a child by delicately taking hold of her chin. When the
horse had again calmed, the gag bit was removed and the horse
was rewarded with a drink and a handful of hay. Thus was
Cruiser the crazy horse tamed.

Word of Rarey's achievement spread and his quota of five
hundred was quickly met. He was hired to teach cavalry officers and riding masters. London cabmen heard his lectures on
how to treat horses humanely. Even the prime minister, Lord
Palmerston, sought a word with Rarey, and Queen Victoria
had Rarey tame one of her surliest mounts. A contributor to
the *Illustrated London News* in 1858 predicted of Rarey that

"His name will rank among the great social reformers of the nineteenth century."

That, of course, never happened. His books, *The Modern Art of Taming Wild Horses* and *The Farmer's Friend: Containing Rarey's Horse Secret, With Other Valuable Receipts and Information*, were soon forgotten.

In the nineteenth century it was common practice, and still is today in some areas, to "fire," or soundly whip, a horse to make him frisky before a sale. One observer remembers seeing "a poor brute, stone blind, exquisitely shaped, and showing all the marks of high blood, whom I saw unmercifully cut with a whip a quarter of an hour before the sale, to bring her to the use of her stiffened limbs, *while the tears were trickling down her cheeks*." The italics were part of the original text and suggest a stretching of the truth: a horse may shed tears from dirt in the eye but, unlike humans, does not weep. Countless horses, though, have had cause to.

In West Africa earlier this century, riders of one tribe rode bareback. To create a safe seat, an historian noted, the horse's own blood was made to act as a glue: "They cut a strip from the centre of the beast's back, about eight inches long by a couple of inches wide, and on this raw and bleeding surface the horseman takes his seat . . . The sore is freshened up with a knife whenever the owner means to start off for a ride." In another tribe, the cruel practice was at least shared by horse and rider: both were cut.

In the Philippines, gaming continues to pit stallion against stallion. Two sex-starved horses are brought into a large outdoor ring where a mare awaits them; the two gladiators may fight viciously over her for an hour, when, typically, the loser flees the ring. The winner then mounts the mare. These contests draw hundreds of onlookers, who bet considerable sums on the outcome.

In 1992, I spent a week at Spruce Meadows, that sprawling show-jumping venue outside Calgary where the world's best riders compete. And I remember in particular a British-born rider with the Swiss team, Lesley McNaught-Mandli. Her horse, Panok Pirol B, failed to negotiate a certain fence — failed, that is, not by knocking the rail down but by putting on the brakes at the very moment that liftoff should have occurred. When a horse balks this way, a little tension grips the crowd, for we know that the rider will almost certainly try the fence again. Cannot let the horse get away with that, the rider thinks, must end on a positive note.

Back to the troublesome fence went McNaught-Mandli. Same result. She managed to stay in the saddle, but the horse did not go gently into that good in-gate. Panok performed a choppy trot that took him up and down, up and down, but hardly forward. The horse issued defiance; his rider, only calm. She had no crop, just her own hard-earned horse sense and her sense of this one horse. Resolutely, she waited — we all waited — for the tantrum to subside, and when it did she patted her mount on the neck and sought the barn. All this took three, four minutes. A small victory had been won, and the applause that rained down on the rider signified our respect.

Why did that moment stay with me? Because what the rider did was not the norm. Most riders carry crops and use them on recalcitrant horses. I know the sound — short, sharp cracks — and hear it often in show rings.

Trainers, called "butchers" in backstretch parlance, still run injured and unfit horses. Jockeys, called "stick riders" or "machine riders," still use the crop freely, or even illegal electric buzzers. Cowboys still wear nasty-looking spurs and sack out young horses. The *palio*, the traditional no-holds-barred horse race through the cobbled streets of Florence and Siena in Italy, still takes a terrible toll in riders' and horses' lives.

A rider struggling to contain an explosive horse still catches the eye in a way that a fine and subtle dressage display does not. One, it seems, is about fire; the other, about water. All those statues of generals on horses pawing the air, of coursers on charging grays, have left their mark. The horse, especially the spirited horse, makes the rider feel tall. "Many of us," wrote Vladimir Littauer, a prolific author and eminent riding instructor, "still derive sincere pleasure out of a brutal mastery of the horse."

The history of the horse-human connection is the history of lessons learned, then forgotten. The kind word seemed always to bow to the harsh; the understanding touch, to the whip, the spur, the jagged bit.

Among the great successes in the world of popular fiction in the mid-1990s was a book about people and horses. Nicholas Evans's *The Horse Whisperer* sat on bestseller lists for years. And I think I know why.

Here is the story in a nutshell. A magazine editor from New York, Annie Graves, travels to Montana, where she desperately hopes a "whisperer" can heal her teenaged daughter's horse, Pilgrim, crazed and almost killed after a horrible highway accident. Somehow, Annie intuits, healing the horse is the key to healing her distraught and now partially disabled daughter, even to healing herself.

The whisperer is Tom Booker, a character likely inspired by Tom Dorrance, Ray Hunt, Buck Brannaman and Monty Roberts — all of whom the author interviewed. Dorrance even gets a mention in the novel: "He'd told her the other day about an old man called Dorrance from Wallowa County, Oregon, the best horseman Tom had ever met."

Tom Booker's cracker-barrel wisdom, his physicality and sensitivity, appear to derive from his life with horses — "the most forgiving creatures God ever made." In his youth, Tom trained troubled horses without ever charging a cent. "I don't do it for the people," he explains. "I do it for the horse."

Likewise, the real Tom Dorrance talks, not about people with horse problems, but about horses with people problems. It's a horse-centered way of seeing. Horses, creatures of flight, are extremely sensitive and aware; humans, more inclined to fight, have lost the awareness once essential to survival. The

4.3   Monty Roberts with Shy Boy, the mustang he gentled in the high
      desert of California in 1997.

horse can hear the whisper and has much to teach humans
about listening. Nicholas Evans had the good sense to be
drawn to this territory. Through sales of his book and the film
that followed, he has been richly rewarded.

The new age of horse schooling naturally has a commer-
cial side. Teachers want to be heard, and thus the multitude
of books, manuals, videos and CD-ROMS. If you assembled
these gentlers, there would be horse-inspired camaraderie;

there would also be territorial tension. Sally Swift on "centered riding," Richard Shrake on "resistance-free riding," Tom Dorrance and Ray Hunt and Monty Roberts on speaking the horse's language, Linda Tellington-Jones on the importance of touch — all are different, but as I pondered their books and watched their videos I was struck by what they share.

Many stress the importance of the horse's body language — lowering the neck and making chewing motions, for example. For Roberts this signals trust, and for Tellington-Jones it does, as well, but she also observes that when a horse lowers his head, his neck relaxes and, as a consequence, he relaxes.

Many of these teachers talk about eye contact: how it can signal aggression. To catch a horse in a paddock, say both Tellington-Jones and Shrake, you approach him quietly, talk softly and look at the ground. These horse-wise experts often choose the same metaphor: riding as a dance, with one partner leading and one following. They calm a horse by patting a spot by the withers, right where a mare would nuzzle her foal. The Tellington-Jones method of touching the horse (sometimes vigorously around or inside the nose or mouth and making circles on the horse's face) seems a more elaborate version of something Indian horse breakers practiced a century or more ago.

In August of 1997, I watched Monty Roberts "start" (he loathes the word "break") three young horses at two different demonstrations in Toronto. I knew what to expect when he climbed into a fifty-foot round pen at Sunnybrook, the oldest stable in the city, in a park set deep in the heavily treed valley that meanders through the metropolis. I had seen Monty school a mustang in the desert, had seen film of him working in a round pen, but I had never actually seen him, in the flesh, perform that thirty-minute miracle he can lay claim to.

Dr. William O. Reed had. The dean of North American track veterinarians, and a New York–based surgeon with a Kentucky horse farm of his own called Mare Haven, Dr. Reed has worked during his illustrious career on many of the finest Thoroughbreds in the world — horses such as Ruffian, Northern Dancer and Secretariat. He watched Monty Roberts work in a round pen with an unbroken horse in December 1995 at the annual meeting of the American Association of Equine Practitioners, or horse veterinarians, in Lexington, Kentucky, the virtual hub of the world's Thoroughbred horse industry. He vividly remembers that day. "I thought it was the greatest communication between man and animal I had ever seen," he told me.

In Toronto, Monty, a squarely built man with a linebacker's body (240 pounds on a five-foot-ten-inch frame), began by talking into a microphone outside the ring. His preamble prepared the onlookers — journalists and booksellers this time — for what they were about to see. At heart, it is quite simple.

When he was a teenager, Monty rode in the wilderness of Nevada for weeks at a time, rounding up mustangs for rodeos. On his belly, peering through binoculars, he would spend hours watching a herd. Monty reasoned that if he could learn how horses communicate with one another, he could better communicate with them.

One day he saw an older mare discipline a colt not quite two years old and full of himself. Full of mischief is more like it. Monty had observed that while the stallion protects the herd, the lead mare runs it on a day-to-day basis. If the herd were a school, the stallion would be the principal and the mare would be the vice-principal. Like an unruly pupil, this colt was kicking mares and biting foals. After the fourth infraction, the mare had had enough. She flew at him, knocked him down once, then again, and finally drove him from the herd.

The isolated colt's every attempt to rejoin the herd was repulsed and he clearly felt panic. When he was eventually let back in, the mare fussed over him, grooming him extensively. When he reoffended he was again expelled, and pretty soon he would leave the herd of his own volition, like a child retreating to his room after a temper tantrum. By the end of four days, the colt was agreeably grooming so many other horses he had almost become a nuisance.

The language of horses, Monty learned, is a body language — "primitive, precise and easy to read." When the mare squared her body and stared at the colt, that meant "Keep out!" When the colt made chewing motions and lowered his head, seeking forgiveness, that meant "Let me in!" Monty calls the language "Equus," and his singular contribution has been to codify its grammar: to tell us that when a horse does *this* with his body it means something quite specific, and when a human does *that* with his body in response it, too, conveys a particular message to the horse.

It is a silent communication between human and horse, and though Monty wears a lapel microphone during demonstrations so he can explain as he goes, the horse pays little heed to his voice (which remains, I should add, even and calm). Equus is mostly about eye contact and body angles. After starting ten thousand horses this way, Monty enters the ring with confidence. It *always* works. But he is also a showman, and part of the drama of these demonstrations derives from the horse — all that power, all that unpredictability. In his preamble, Monty accentuates the risk.

Outside, that overcast day in Toronto, stood frisky proof. A fifteen-hand chestnut named Happy Go Lucky, a Trakehner-Thoroughbred cross, anxiously eyed the people milling nearby. "He's a little stubborn," his handler told me, "but he's very manageable."

In the round pen, Monty introduced himself to Happy Go Lucky, and rubbed the horse's forehead often as he talked. Monty said he aimed to accomplish in thirty minutes what traditional horse breaking requires five to six weeks to do: introduce the horse to bridle, saddle and rider. He flicked a light cotton line at the horse to send him round and round the perimeter. "Don't go away a little," he told the horse. "Go away a lot." We got a sense of the gelding's speed and power as he kicked up the turf on his turns. Onlookers were peering through a grid-metal fence, only feet away. We could feel the rush of wind as he passed, see his wide eyes, hear his loud breathing.

But this horse was indeed manageable. He soon learned to respect, but not fear, the tossed line. Monty, in essence, put the horse to work by making him trot around the pen.

The alternative to work, Happy Go Lucky soon gathered, was conversation. Monty gave us what he imagined were the horse's thoughts — that "this man with the belly" seemed not so bad after all. The key "signs" then came, one after the other: the horse, still circling, locked one ear on Monty (the other ear continued monitoring the crowd), then stuck out his tongue and made chewing motions; finally, he lowered his head until it was inches from the ground. Like the colt in the high desert to the mare, he was saying, "Can we talk?"

Monty then angled his shoulders away from the horse, even turned his back to him. Happy Go Lucky, still somewhat skeptical, delicately approached. The horse was alone and thus fearful, for to be alone is to be exposed to predators. (Observers of wild-horse herds point out that stallions who lose their harems and must live alone quickly die. The solitude invites stress, disease in turn, then death. This is only one measure of the horse's powerful herd instinct, an instinct that Monty harnesses in the ring.)

Happy Go Lucky wanted to join a herd. By this point, Monty's herd looked a good bet, since Monty had done nothing to harm him. In fact, getting stroked on the forehead began to seem a better choice than the "work" of all those trips around the perimeter. Slowly, the horse stepped toward Monty's back and nuzzled him, even followed Monty as he turned left, then right. Monty calls this "join-up."

The subsequent bridling seemed to me astonishingly easy: how would you respond to a stranger putting a leather contraption on your head with a metal piece that fits inside your mouth? The saddle was also accepted with good grace, and *then* came some serious bucking. Monty looked on, unconcerned. Later, when a TV camera clacked, Happy Go Lucky spooked and instinctively sought Monty. In that sudden storm Monty was the nearest port.

When a young rider eased his body across Happy Go Lucky's saddle, much as Indian horse breakers used to do, the horse abided this. Neither did he complain when the rider got a leg over and put both feet in the stirrups. "In traditional schooling, the horse bucks the rider 95 percent of the time," Monty told us. "Ninety-five percent of horses started the way we have today do *not* buck. We've turned the stats around."

Monty knows when to make eye contact with an unbroken horse and when not to, where to touch the horse first, whether to move slowly or quickly. All this he knows because he has learned his equine manners and grammar. He rejects the label "horse whisperer" as distasteful and inaccurate. The term implies somehow that his skill owes something to his "touch" or to his Cherokee ancestry. Monty Roberts is no whisperer; he is a teacher. And in time, he says, we could all learn what he knows.

The next night, at the York Equestrian Centre north of Toronto, Monty did the same thing with two other horses, the second a princely and fiery dark gray colt by Abdullah, a Trakehner stallion who won both gold and silver medals in show jumping for the U.S. at the 1984 Olympics and two world championships after that. The huge arena drew one thousand people, the biggest crowd in all the years that Monty Roberts has done demonstrations.

Well into the evening, Monty offered a little lesson in the importance of eye contact with the two-year-old pureblood Trakehner called Abdullar. He told us what would happen and where, and he was as good as his word. As Abdullar circled past an appointed spot on the perimeter, Monty dropped his gaze to the horse's hips and the horse actually stopped. Stopped dead. When Monty quickly returned his eyes to the horse's eyes, the horse picked up the canter again. There was a little gasp from the audience. "Eyes on eyes, he flees," said

Monty. "Eyes on hips, he stops. I'm here to tell you that a horse can see your eyes from half a mile away."

I was left doubly impressed. First, I was struck by the intelligence of these horses. You could almost see their minds working as they made conscious decisions about when or if to approach this stranger. I also sensed how powerful was the horse's language and how strong the compulsion to be social. "His own language is taking over," Monty told us, "and he's *got* to comply."

Not shy about extolling his achievement, Monty reminded his audience of what could go wrong. A young horse might eye his first rider, Monty said, and believe him to be a predator. True enough, but the predator in this case was a laid-back, gum-chewing, black-hatted Idaho cowboy. Still, the horse is wary, and the gentler had better be also. In New York only days before, a particularly feisty horse had tried to kick Monty innumerable times, and had actually bitten him three times. Monty's son Marty, who was in the audience that night, later told me he began to fear for his father's safety. But Monty persisted and eventually gentled the horse, and the Bronx audience rewarded him with a standing ovation. We did the same in Toronto.

Three green horses brought to bridle, saddle and rider in ninety minutes. Not a harsh word spoken, not a hand raised in anger. That other Monte, who "rode down the gray," would have been impressed.

"Hurt the horse," Marvin E. Roberts told his young son Monty, "before the horse hurts you." Where did that fear of horses come from? Maybe from stories like this one.

Cowboy accounts of roping mustang stallions may have been overstated, but the spirit of truth in them leaves a chill. "Probably no more vicious animal lives than a mustang stallion," wrote one nineteenth-century mustanger named Frank Collinson. "He bites, strikes with his forefeet, kicks, tries to jump on you or kill you in any way he can. He is the most dangerous animal I have ever had anything to do with."

Collinson describes how he tried to capture a big sorrel stallion, how the roped horse turned on him and severely bit his leg. He vividly remembers the sound of the stallion's teeth when they slid off his thigh, clacking "like a sprung steel trap." The horse was not dissuaded by a warning pistol shot but used the rope to jerk his assailant's horse to the ground, then rushed him at full gallop — "with his mouth open and ears set back and eyes like balls of fire." Bullets from a friend's Winchester felled the stallion, but only a few feet from the man standing with his own mouth open and his boot full of blood.

While Margaret Cabell Self was penning her quite sympathetic book on horses, Marvin E. Roberts was writing his. *Horse and Horsemen Training* displays a black-and-white photograph of the necessary bits and halters and ropes, the latter, though new, clearly stained with blood.

For the bucking horse, urged Roberts Sr., fill a burlap sack with tin cans and attach it to his saddle — a little terror to let the bronc know who's boss. When a horse refuses to leave the herd (called "herd-bound"), "put your left hand on the saddle horn and hit him on the top of the neck right up between the ears as hard as you can." Where to hit, what to use and when ("some colts have to be hit more than others") are all in the little blue book.

Roberts Sr. professed to love his horses, and in his own eyes he did. His thoughts on curbing and disciplining horses seem in keeping with the time. No one branded him cruel. On the contrary, for the gift of his horsemanship (he taught many local children to ride), the grateful citizens of Monterey, California, named a rodeo arena after him.

Traditional "sacking out" was as much war on the horse's mind as on his body. The aim was to sap his resistance. Monty would watch his father take three weeks to break six horses. They were haltered and tied with rope to posts laid thirty feet apart around the corral. Then weighted sacks or tarpaulins were fitted over them and tightened with more heavy rope. The horses fought this, of course, bloodying and often hurting themselves.

In the wild the mountain lion often preyed on mustangs. The cat would leap from a tree onto the horse's back, clinging with its front paws and attempting to disembowel the horse by raking his flank. Some horses managed to buck the lion off but often bore the scars; any bronc-buster seeing such scars on a horse wisely gave him a pass — hence the expression to "pass the buck." The nineteenth-century cowboys who witnessed the ferocious bucking by mustangs whenever *anything* was put on their backs guessed that the fear was inherited. Thus, a weighted sack or saddle inspired in a horse an ancient and deep-seated terror.

Though some horse breakers were cruel, just as some people are, the cowboys I have met are anything but. Ranchers likely saw sacking out as the only option, and an efficient one at that. Cutting horses work so hard that their working lives are correspondingly short. Ranches, then, needed great numbers of horses for cattle drives and general ranch duty. Since gentling had no champion, the rough methods of the bronc-busters — professionals who went from ranch to ranch charging $5 a head — prevailed.

As part of sacking out, one hind leg was tied up with ropes and connected to a rope collar around the horse's chest. With the horse's foot off the ground, bucking was impossible. More sacking out ensued, and finally a saddle was put on. The end result may well have been a cooperative horse, but something precious, Monty Roberts came to believe, had been lost.

Great trainers, such as Ian Millar in the show-jumping world, talk about a horse's generosity. It is a great gift often wasted. Millar once had a horse, a fine jumper prone to nervousness and pacing in his stall. A previous owner had been too free with the crop and the well of generosity was not as deep as it might have been.

Generosity means that the horse will do something extra,

but only for you, his trainer and rider. "It never ceases to amaze me what a horse can give," says Jim Elder, another Canadian show jumper and an Olympic champion. "You'll be in a tight corner or moving at speed, and you think — it's so difficult the horse can't do this next jump. But he does!"

Monty Roberts calls that quality "willingness," and he argues that to destroy it by making the horse work out of fear is both senseless and unforgivable.

Horse trainers will say that the horse is a wonderful teacher, that horses never lie, that even the malingerers are telling a truth. You might think such trainers mean something quite mystical, and perhaps they do. But the handler must show perseverance and trust to get back from the horse anything like willingness or generosity.

What better place, then, for the horse to teach these virtues, than in prisons? Over the past nine years, more than four thousand wild horses have been gentled in Wild Horse Inmate Programs in four American states (California, Colorado and Wyoming; New Mexico's has ended.) The Bureau of Land Management culls wild-horse herds before putting individual horses up for adoption, and sometimes the job of training them falls to convicts — most of whom have never seen a horse, except on television.

This pairing of horses and convicts offers practical skills that the prisoner might use when he is free, along with lessons in humanity that might just *keep* him free.

Tom Chenoweth, a horse trainer with more than thirty years of experience, oversees an inmate program in Susanville, California. Prisoners, he says, have no choice but to go slow with a wild horse whose first experience of a human — being dragged into the prison yard on the end of a rope — has been brutally imprinted on his brain.

"Among inmates there is no kind word spoken," Chenoweth told me, but those manners won't work with wild horses. "You cannot bully a horse or make him do anything he doesn't want to do. A wild horse doesn't even want to be patted. He has to be taught that it's pleasurable. You have to honestly and openly communicate with the horse." It would be ludicrous, he said, to suggest that contact with horses transforms men who may lack morals or ethics. What is true is that patience and effort are rewarded.

It may take weeks, even months before the horse lets a human close enough for grooming, haltering, bridling and saddling. The injuries that sometimes occur offer lessons in respect and the patience required of prisoners is extraordinary — a single incident of lost temper can undo weeks of hard work. But at the end of that long and delicate process is a fine horse. And, perhaps, a finer human.

Today, the grand old man of horse gentling is Tom Dorrance. Tom's brother Bill, now in his mid-nineties, is also a great horseman and was long a mentor to Monty Roberts, especially when Monty was a young man. No one had any time then — this was the 1940s — for Monty's ideas about understanding horses. Bill Dorrance did, and so the world of horses owes a great deal to the Dorrance family.

Tom Dorrance, now in his late eighties, began riding a cutting horse and moving cattle on his father's ranch in Oregon when he was only five years old. He was the second youngest of eight children, and sometimes he and his brothers would ride young horses for neighbors who would later ask how the horse went. "Well, Tom's riding him," Bill might reply. If a little ol' kid could ride him, the thinking went, the horse was probably alright.

"Tom was easy with the horses," Bill once wrote, "and they all worked for him. He wanted to get along with them. As time went on Tom figured out how to get a relaxed feel with horses . . . That relaxed feel really felt good to Tom and the horse." It was in Tom's nature to live peaceably with all creatures — human and animal. A clue to his character may lie in the book Tom later recommended to riders: *A Kinship With All Life*.

Tom was a small man — five feet six inches and for the first thirty years of his life he never weighed more than 130 pounds. "I couldn't manhandle a horse," he says. "I was often alone and far from home. If he got away, I'd have to walk." So Tom Dorrance, on his own and still very young, tried a different approach. It involved winning the horse's trust and seeking to understand what the horse meant when he whinnied or lowered his head or nudged Tom with his head. Like Monty Roberts, he spent hours watching horses in the wild and in the paddock. Without ever reading Xenophon (he quit school before finishing grade 8), Tom came around to Xenophon's way of thinking.

Tom Dorrance thought he would try to see the world through the horse's eyes. Some cowboys today call him "the horse's lawyer" or "the patron saint of horses."

Eventually, Dorrance wrote *True Unity: Willing Communication Between Horse and Human*. It's about "reading" a horse and respecting the horse as an individual. One of Tom's students describes in that book what he learned from the old horsemaster. The man admitted to a foul temper and using it on his horses; Tom turned his life around. "He probably made me a better individual from understanding horses," the man said. "He taught me to *see* what I look at . . . in everything — horses, cattle, people."

Along the way, Dorrance has taught a great many other students, who continue to pass on the art of gentling. One of them is Ray Hunt, a craggy-faced man in his seventies who still goes on the road demonstrating how to school a green horse. In similar clinics at a six-thousand-acre ranch, the Dead Horse, near Santa Fe, New Mexico, he does in a matter of hours what bronc-busters took a week to do or sometimes failed utterly to accomplish. Hunt also wrote a book, *Think Harmony With Horses*, and he dedicated it to the old master Tom Dorrance. "The slower you do it," Hunt writes, "the quicker you'll find it."

What distinguishes "natural horsemanship" from the old

way is not just the gentleness of one and the violence of the other. It's pace. Monty Roberts has a good line that echoes Hunt's: "Act like you've only got fifteen minutes and it'll take you all day. Act like you've got all day, and it'll only take you fifteen minutes." Dennis Reis, another gentler, says, "You've got to stop thinking in people time and start thinking in horse time."

Today, it seems there *is* no time, but many smart and skilled horse people are saying that to train a horse well you must take time. The horse, in effect, asks you to slow down. And that message is refreshing and welcome news for all who love horses.

Tom Dorrance urges riders and trainers never to let frustration take hold. Walk away from the horse, cool off, then come back. It is a lesson won from a life with horses, and many cowboys know it.

While I was riding in Wyoming, Bruce, a young wrangler, got a truck stuck — well and truly mired — in a mountain creek. As we rode into camp at day's end Bruce approached Skip, our guide and his boss, and uttered the hope that Skip had had a really *good* day, because it was Bruce's unpleasant duty to show him evidence of his own really *bad* day. Bruce led him over to the truck, which Skip then freed, ingeniously, by using blocks of wood as levers and by nudging it from behind with another truck when pulling failed. Skip's calm admonition to the wrangler later on was that moving too quickly had compounded matters: "You shoulda just sat down for ten minutes. Had a cup of coffee and thought things out." What works with horses works with horse-powered vehicles.

Ray Hunt also knows about taking your time. He teaches a horse to back up, for example, by a subtle shift in weight to the rear and rewarding even the tiniest effort on the horse's part to respond. It's not about yanking or fighting the horse; it's about rider and horse becoming a single unit. And that takes time.

4.4   Ray Hunt, horse gentler: it takes time for horse and rider to become a single unit.

Marvin E. Roberts's advice to cure the "barn-sour" horse (one who never wants to leave the barn) was to hit him hard. Hunt's advice for the same problem is very different and actually quite simple. It's one of his mantras: "Make the wrong things difficult and the right things easy." Make hanging around the barn problematic for the horse, simply by keeping his feet moving. Do not let him park, and when he makes even a small move away from the barn, reward him.

The centaur of Greek mythology has always been the symbol of the perfect bond between horse and rider, and that is what Ray Hunt aims for. Which makes Xenophon of ancient Greece and these old-guy gentlers of today brothers of a sort.

It is hard to say how many horse trainers are moving in the direction of the gentlers. Some show jumpers still "rap" their horses' knees in training sessions to make them leap fences. The "wild horse roundup," in which teams of three men strive to saddle, bridle and ride completely terrorized wild horses, is still a feature of many rodeos.

4.5   The wild horse roundup, still a cruel feature of many modern rodeos.

But my sense, from talking to dozens of horse trainers, is that the motion — not just in horse circles, but in society in general — is toward a more benevolent attitude to the animals in our care, coupled with a growing sense of fascination with their inner lives.

During the brief span of time in which humans and horses have interacted, the response on the part of the human has always been either utilitarian (How can I exploit this animal?) or egomaniacal (Don't I look grand on this fine gray horse?). Few humans have ever wondered what animals were thinking or feeling, or even granted the possibility of animal thought or emotions. Only in recent decades have doctors ceased believing that newborn infants neither feel nor recall pain and thus have no need for anesthesia during invasive procedures. And if we have dismissed babies as sentient beings, then doubly have we dismissed animals.

For much of human history, a strict hierarchy put God at

the top, humankind below, and animals near the bottom. Man, says the Book of Genesis, would have "dominion over . . . every creeping creature that moveth upon the earth." Most western cultures gave little thought to whether the animals they ate also remembered, felt, feared or sorrowed.

Many of us no longer think that way. The rising tide of vegetarianism, especially among young people, suggests a startling new awareness of animal consciousness. In academic circles, many thinkers — among them the anthropologist Elizabeth Atwood Lawrence and the entomologist Edmund O. Wilson, the only two-time winner of the Pulitzer Prize — have advanced a philosophy they call "biophilia." It comes from the Greek words *bio* (meaning "life") and *phil* (meaning "loving"). Humans, they argue, have an innate tendency, however repressed, to focus on other life forms.

Wilson and Lawrence further argue that as we come to understand other organisms — how horses interact in a pasture, the various tasks of a line of ants in the jungle, the mood of your dog walking ahead of you at dawn — we value them, and ourselves, more. *Controlling* nature may yet yield to *observing* nature.

Monty Roberts, in *The Man Who Listens to Horses*, describes a fellow horse trainer named Greg Ward. Horses from the Ward ranch in California have won twelve world championships in cow horse competition and millions of dollars in the show ring. These horses are clearly of the highest quality. What is astonishing, at least to trainers of the old school, is how Ward eases young horses into their working lives: for the first twenty days that a rider is on the horse's back, the horse calls *all* the shots. If the horse wants to graze, trot, even roll, the wish is granted. There is no chance for resentments or neuroses to form. The method is rooted in affection for the horse, but clearly, Ward also believes this way is more efficient than the old way.

Roberts calls it "a new beginning in the relationship between man and horse."

———

Horse trainers may never agree on the philosophy of teaching skills to a horse, but on this they might agree: schooling a horse is about paying attention to particulars. Find the key to that one horse.

About 340 B.C., a young Macedonian prince named Alexander was looking on as his father, King Philip, was considering whether to purchase a huge black stallion. No one could even mount him, such a fury was this horse, and his own grooms stayed well clear of him. The price being asked for the horse was high, so clearly the owner thought him a superior animal.

The king was angry that such an impossible creature had been presented to the royal family, and he ordered the horse taken away. But amid all the prancing and pawing of the great black horse, the twelve-year-old prince had noticed something. "What a horse they are losing," he said with some sadness, while regretting aloud the handlers' lack of skill.

The king, like many a parent, said as much as "So you think you can do better?" The prince was certain he could. He offered to pay the full price of the seemingly intractable horse if he was unable to tame him. Historians suggest that the price being asked was thirteen talents, or more than $10,000 in modern currency — a lot of money, even for a prince. The pride on the line was incalculable.

Alexander approached the horse, took him by the bridle and simply turned him to the sun. This was the key, for the black horse had come to fear his own shadow. Alexander spoke softly to him, stroked him, then mounted him. They took off in a gallop, which alarmed the king, but soon they were back. The unruly horse had met his rider.

The king wept with joy. At that moment he realized with certainty that his son was destined for greatness. "Macedonia," he said, "is too small for thee." The king was right on that score: Alexander the Great would conquer much of the known world.

No one else ever rode the horse. Neither Alexander nor the horse would have acquiesced in any case. Plutarch reports that "In Uxia, once, Alexander lost him, and issued an edict that he would kill every man in the country unless he was brought back — as he promptly was."

Alexander called his war horse, black with a white star, Bucephalus, or Ox-Head. An artist once drew a portrait of Alexander and Bucephalus. The royal rider turned up his nose at it, but the royal horse neighed a greeting to his likeness. Rather bravely, the painter suggested that "Your majesty's horse is a better judge of a painting than your majesty." Other kings might have called for the man's head, but apparently the painter kept his.

Bucephalus would kneel before his master to be mounted, and into battle they would go. After many campaigns the horse died at the age of thirty, an unusual life span for a horse in those days. During the final battle in India, horse and general waded into the fray, and the horse took spears in his neck and flank but still managed to turn and bring the king to safety before dying.

Alexander was overcome with grief, and later named a city after Bucephalus. "He was as dear to his master," wrote one historian, "as Alexander was terrible to the barbarians." Some historians suggest that the ancients typically saw horses as weapons in war, took no particular pleasure in riding and felt no real affection for their mounts. If so, Bucephalus was clearly an exception. All this because a young boy, a gentler before his time, listened to the horse.

CHAPTER 5

THE HORSE IN BATTLE

After God, we owed the victory to the horses.
RECORDS OF THE CONQUISTADORS

Four things greater than all things are —
Women and Horses and Power and War.
RUDYARD KIPLING

THE PAINTING IS called *Scotland Forever! The Charge of the Scots Greys at Waterloo, 1881*, by Lady Elizabeth Butler (1846–1933), and it captures in a fierce, exaggerated way the tumult, menace and pure glory often attached to the notion of war.

The six-foot-wide oil on canvas is filled by a line of fur-hatted hussars brandishing sabers and riding charging white horses. The silent, crushing wave of cavalry, the demented look in the horses' eyes, those flying manes, all coalesce. The painting bids you pause, but the longer you stare the greater the desire to step back or to duck its line of energy.

5.1   *The Charge of the Scots Greys at Waterloo*, 1881: Glorious art,
sobering reality.

Other paintings, of Napoleon, say, depict the emperor on a muscled and taut gray with black mane, the reins dropped confidently over the saddle. Napoleon is using his looking glass to monitor the battlefield on the distant plain below and is handing off, with his left hand, a map to an underling. Another famous painting, by Jacques Louis David (1748–1825), *Bonaparte Crossing the Alps, 1800,* has the mighty gray pawing the air, the emperor caped and invincible looking, one finger pointing the way to victory. So impeccably groomed is the horse, so gloved and elegant the emperor, that both could be going to church, not to war.

But read the literature on the horse in war and a different picture emerges. The Battle of Waterloo involved thirty thousand horses in a veritable meat grinder. The paintings lie: horses seldom looked glorious before battle, and certainly not after.

One photograph in Charles Trench's *The History of Horsemanship* sticks in my mind: a disconsolate soldier atop a mangy, crusted, woebegone horse en route to the Crimean War. The horse looks small, almost ponylike, and sad. Trench quotes the wife of one officer who watched the troopers' return and who remarked on what "a piteous sight it was — men on foot driving and goading the wretched, wretched horses . . . a cruel parade of death."

A British officer in the Crimean War, Lord Cardigan, ordered the Charge of the Light Brigade on October 25, 1854. He liked to move forward at a gallop. This habit alone killed or disabled almost half the horses, who were grossly underfed prior to that suicidal engagement at Balaclava. Though plenty of food was stored six miles away, Cardigan would not allow his mighty steeds to be used as lowly pack horses (of which there were none). So he let them starve. They died after gnawing leather straps, ropes and one another's tails to stumps.

The battles themselves were hideous. Here is a scene from the Charge of the Light Brigade as told by the Marquess of Anglesey, a British historian: "The old grey mare of Trooper

John Lee of the 17th, who was killed, kept alongside its neighbour for some distance, all the while 'treading on and tearing out her entrails as she galloped, till at length she dropped with a strange shriek.'" Sergeant Talbot of the same regiment "had his head clean carried off by a round shot, yet for about thirty yards further the headless body kept the saddle, the lance at the charge, firmly gripped under the right arm."

The reality looked nothing like Lady Butler's painting. Unhorsed in the charge at Balaclava, about three hundred cavalrymen were now "struggling along, some crawling, some limping, others running, some dragging loved horses bleeding to death behind them." Some 475 horses were killed that day (including forty-three later shot as unserviceable owing to wounds). On the battlefield, riderless English horses were rounded up by Cossacks and offered to the highest bidder.

Worse than the battles in a way was the horsemanship. Trench tells the story of a British mounted infantryman who asked his commanding officer during the Boer War expedition in South Africa at the turn of the century what he should feed his horse, mutton or beef.

During the Boer War, professional soldiers in the Royal Artillery did care for their mounts well, for they knew something of horses. But other elements of the British cavalry knew next to nothing: three-quarters of the enlisted men in the Imperial Yeomanry had never sat a horse before passing "a riding test," and the Mounted Infantry had to learn on the job. Rudyard Kipling had a phrase for it: "three days 'to learn equitation' and six months of blooming well trot."

Only the South African Colonials, who rode their own horses, had the sense to graze and water them at every opportunity. Keeping a single, overloaded and overworked horse sound and fit during long marches required great skill, which most of these men lacked.

One general, who claimed he alone tried to stop the waste, railed about it. "I never saw such a shameful abuse of horseflesh in the whole course of my life as existed throughout the whole campaign, and not an attempt was made to check it . . . I was shocked, I was horrified."

The numbers tell the tale. British forces in South Africa lost 7 to 8 percent of their horses *every month*. Of the 494,181 horses sent, 326,000 perished — only a small number in actual fighting. And lest you think that the South African campaign was an isolated case or specific to the British, consider the Prussian army in the Franco-Prussian War of 1870. "Operating," as Trench notes, "in a temperate climate and a fertile country, close to their bases and well served by rail communications, they lost over a million horses in eight months."

"War is hell," said General William Tecumseh Sherman. He fought for the North in the American Civil War, and when he uttered those words to graduating cadets at a Michigan military academy in 1879, he was speaking from his own dark experience. "It is only those who have neither fired a shot nor heard the shrieks and groans of the wounded," he said, "who cry aloud for blood, more vengeance, more desolation."

But if war has been hell for humans, it has been the same for horses. What made the horse such a terrible weapon in war was not just his speed and power but his willingness to ride into the very hell that General Sherman decried; to do what was asked of him, despite his rider's screams, the clash of metal on metal, the piercing sound of guns.

The horse was a willing warrior, and no other animal has been so consistently or widely enlisted to fight men's wars. The elephant, the camel and the dog saw only spot duty. For most of the six thousand or so years of equestrian history, horses somewhere on the planet were being ridden to war. Luckless barbarians fleeing Roman cavalry officers were menaced by spears, swords *and* the teeth and hooves of Roman horses. In desert battle, Arab horses fought one another with the same fury that their riders did. Soldiers throughout human history could not have asked for a more generous or devoted ally than the horse.

Like war itself, the war horse chronicle — with all its tales of courage and folly and pride — stirs up a lasting melancholy, in part because of the great fondness that certain soldiers felt for their doomed mounts.

Lieutenant Edward S. Godfrey, who rode with Custer towards the Little Big Horn but split off beforehand, wrote in his diary along the way that at lunch time, "When the haversacks were opened, the horses usually stopped grazing and put their noses near their riders' faces and asked very plainly to share the hardtack . . . The old soldier was generally willing to share with his beast." During the American Revolution, a patriot named Samuel Dexter was so grateful to his horse that he established a retirement fund to ensure the horse's care. Some war horses were buried with full military honors. And Blackfoot warriors, otherwise hardened to deprivation and taught never to complain, wept openly when their horses died.

The melding of horse and armed rider created an enduring effect. The romance that for so long was, and is still, attached to the People of the Horse — to the Comanche and Crow and Sioux of the plains, the Huns and Mongols, the Turkish Mamelukes — helps explain why sword-carrying cavalry continued to charge at guns for centuries longer than they sensibly should have.

What I see in old photographs of cavalrymen and dragoons is unabashed haughtiness. Fully armed, trained to kill, with all that power underneath him, the trooper must have felt almost invincible. Foot soldiers and peasants were cut down by cavalry all through the ages. The carnage that took place, especially at the hands of such feared horse peoples as the Mongols, inspired dark legends. Imagine what terror was induced when great dust clouds appeared on the horizon and got steadily closer until the unmistakable sound of thousands of horses' hooves told all in their path that what they feared had finally come.

Czechs once believed that the devil created horses. Europeans thought Satan traveled on the back of a coal-black horse. The Chinese thought a demon called Horsehead tortured the damned in the next life. And in many folk tales, phantom steeds and headless horses foretell death.

But a mounted soldier is, of course, not invincible and the horse, for all his natural weaponry, is actually a shy warrior. So much of horse etiquette — who grazes where, who waters first, the hierarchy in the pasture — is determined by display. A look, ears back, a swishing of the tail — horses constantly read one another and diplomacy rules their society far more than aggression does.

Mares and geldings skirmish, but only the stallion warms to actual war. In the wild, fights will range for three days or more and the combatants may cover thirty miles in the course of their attacks and retreats. The American wildlife ecologist Joel Berger, who has spent years studying wild horses in Wyoming and elsewhere, reports that 96 percent of adult males show signs of bite-related wounds. One stallion had fifty scars on his body. But deaths from these battles are not common and only one in almost seven hundred observed encounters actually led to a harem takeover. By comparison with human society, equine society is a veritable peace corps.

For all that, horses do make war when men call upon them. And, if our attraction to horses is indeed "bred in the bone," as the proverb says, then the bone is spattered with blood, equine and human both.

The horse was first used in war four thousand years ago. Someone with imagination must have looked upon the donkey-drawn wagon and come up with the tank of its day — the horse-drawn chariot with two men inside, one to drive the horses and one to fire the arrows. Several thousand chariots at a time could charge an enemy and in minutes inflict terrible damage on foot soldiers.

The chariot, though, was useful only on roads and flats. Mountains and rivers seriously impeded its progress, and by 900 B.C. the mounted soldier had supplanted it.

Around 700 B.C. the Huns, Asiatic nomads who also devised the stirrup, invented the right clothing to wear in the saddle — trousers and boots. This marked a vast improvement over bare legs and sandals. First of the so-called horse people, the Huns were the masters of hit-and-run warfare. Wearing no armor and using only bows (warriors were particularly adept at galloping forward, turning in the saddle and firing up to six arrows a minute at enemies behind them), they completely outmatched their Chinese opponents, who stuck with the short sword and the chariot. "Their country," wrote one Chinese historian, who both dreaded the Huns and admired their horsemanship, "is the back of a horse." In the saddle, the Hun ate, drank, slept.

Even when the Chinese taught themselves to ride and shoot like Huns, the enemy still confounded them. The formidable Great Wall of China was built in part to keep the Huns and their horses out. The Huns then looked west, and by A.D. 450 they had conquered most of what is now Russia, Germany and Poland and had invaded Rome. They might have continued had Attila not died, freakishly, of a nosebleed.

The world cowered before these terrible centaurs. "Swift as the wind," one Hungarian chronicler wrote, "the riders came up like a tornado and disappeared like a flock of birds."

Hun cavalry could cover ninety miles a day. Each rider took a string of four or more horses on conquering journeys that sometimes spanned thousands of miles. When one horse showed signs of tiring, the rider mounted a fresh one. In time of need, the horseman would make a delicate cut in the neck of his horse and drink a cup of blood without harming the animal. The Hun warrior was as kind to his horse as he was unmerciful to his enemies, and the reward for old war horses was retirement to a good pasture.

As fierce as the Huns were, the Mongols who came after them seemed even more terrible and more invincible than their steppe ancestors. During the twelfth and thirteenth centuries, Mongol horsemen dominated the known world the way no

5.2   Mongol archer: Brilliant horsemen who conquered much of the known
       world in the 12th and 13th centuries.

other people had before or have since. They overran northern
China, Korea, Tibet, Central Asia, parts of Russia, northern
India, Poland and Hungary and prepared to raid Vienna and
Venice, retreating only at the news that Genghis Khan's son
had died.

Mongolian babies, it was said, were bathed daily in cold
water to harden them, and by the age of sixteen Mongolian
boys were brilliant horsemen. They could ride two days and
two nights, sleeping, eating, even relieving themselves, in the
saddle.

Historians debate the number of people slaughtered by
Genghis Khan and his warriors. One estimate puts it at more
than eighteen million. On the other hand, some cities — Bok-
hara and Samarkand, for example — simply capitulated at the

very appearance of Mongol horsemen, such was their reputation as unbeatable in war. What is known is that Mongol chiefs gathered intelligence first and attacked second, that they fought diversionary battles with the enemy to mask the thrust of the attack and that the *yasa*, or Mongol code of law, meant instant death to any warrior who abandoned a comrade.

John Keegan, the eminent British historian and author of *A History of Warfare*, called the Mongols "warriors for war's sake, for the loot it brought, the risks, the thrills, the animal satisfactions of triumph." A latter-day biographer of Genghis Khan quotes him as replying thus to the suggestion that falconry was the greatest pleasure known to man: "You are mistaken. Man's greatest good fortune is to chase and defeat his enemy, seize his total possessions, leave his married women weeping and wailing, ride his gelding [and] use the bodies of his women as a nightshirt and support."

The Mongols' customary food — milk, meat and blood — was derived from one source: the horse. Blood was drawn from the veins of living horses, stored in gut bags, thickened over a fire and then fried like a black pudding.

The Hun and Mongol invasions rocked Europe, as did the Crusades. An army of mounted Christian knights from Europe took Jerusalem from the Muslims in 1099 and for much of the twelfth century fought a seesaw battle against Saladin. Armored knights of the Middle Ages were as slow and heavy as their Muslim enemies were light and fast. Where a modern jockey on a Thoroughbred can do forty miles an hour, a knight could manage about fifteen. He rode a lighter horse (the palfrey) to the site of the battle and a heavier horse (the destrier, about the size of a draft horse) to charge the enemy (thus the expression "to get on your high horse"). Assuming the knight weighed 176 pounds and his armor seventy-nine pounds, the further weight of saddle, stirrups and other tack meant the poor horse had to bear a total weight of 448 pounds.

The Arabs' horses — "fast, spirited and elegant beast[s], pampered and often hand-fed," says Keegan — were effective

5.3   Female knight on her high horse: the Crusades pitted Christian against
       Muslim, power against speed.

in battle, but there were precious few of them. Somewhat like
his Christian enemy, the Arab rode his camel to war and
saved his sleek and precious horse until the moment he drew
his scimitar.

The exorbitantly rich Bedouin vocabulary pertaining to horses is a measure of how important the horse was in that culture. There are thirteen Bedouin words for herds of horses; one hundred for horse colors; twenty for a noble, high-bred horse or mare; eighteen for a fiery horse; fifteen for starting a horse; fourteen for trotting; twelve for prancing; and twenty-one for different kinds of walk.

The Arabs, Keegan contends, were not horse people in the way that the nomadic Hun and Mongol raiders were. They were architects and builders, patrons of art and literature, and in the jihad, or holy war, that would overwhelm North Africa, the Middle East and what is now Turkey, they would eventually hire Turk mercenaries — horse people of the steppes riding shaggy little ponies. The Mamelukes were blood relations in every way to the Mongols and their mounts.

Legend has it that Turkish women gave birth in the saddle, and it did indeed seem that Turk warriors were born there. No soldiers, not the religiously inspired forces of Islam or the professional soldiers on the Christian side, succeeded in withstanding the fleet and merciless Mongol invaders.

The Mamelukes did.

They were part of an Egyptian army that confronted Tamerlane and his Mongols at Ain Jalut, just north of Jerusalem in 1260. It was an extraordinary battle, still studied by military historians today. Turk against Turk. Horse people against horse people.

Tamerlane dwarfed even Attila and Genghis Khan in his appetite for atrocity. Tales of pyramids of human skulls date from his campaign. The Mamelukes, then, saved both Muslims and Christians from a dark and savage reckoning.

Like the samurai in Japan about this time, the Mamelukes were a warrior class. Taken as slaves, they were taught both the Koran and the *furusiyya* — that special mastery of horse and arms that allowed them to hand the Mongols their first major defeat.

The invention of gunpowder by the Chinese in the 1300s,

followed by guns and cannon, eventually ended the era of the horse soldier. But it was not easy for a proud cavalryman to give up his horse or for his superiors to concede the folly of riding toward spit fire. Yet they persisted, and for a century or two longer than seemed rational. Polish cavalry did *not*, as Nazi propaganda proclaimed, charge German tanks during the early days of the Second World War. But there is some truth in the image: what Keegan calls "the twilight of the war horse" was a sad, slow unfolding.

When Napoleon invaded Egypt in 1798, the Mamelukes' *furusiyya* offered a futile response to muskets and cannons. Napoleon, though, seemed touched by Mameluke gallantry and had a Mameluke named Rustum as his personal servant right to the end of his reign. At Cairo in 1811, the Mamelukes were again slaughtered, this time by the Ottoman Muhammad Ali. Gunmen against swordsmen — even swordsmen on horses — was no contest.

If soldiers were loath to give up the horse as a weapon in war, their reluctance stemmed from confidence in their own equestrian skills, which had grown increasingly elaborate. Cavalry schools in sixteenth-century Europe taught officers and their mounts deft moves meant to heighten their efficiency in battle and enable a hasty retreat.

There was the pesade, in which the horse rears up to protect his rider from bullets; the half pirouette, a quick turn that allows the rider to mow down enemy soldiers; and the piaffe, a high-stepping, in-place trot that essentially kept the equine motor idling while the rider hacked at the infantry with his sword. Advanced horses and their riders also learned the difficult capriole, in which the horse tucks in his front legs and thrusts out his back legs, rising in the air without moving forward.

Several years ago I watched a performance by Andalusian horses at Jerez de la Frontera in Spain. While there is much to admire in the precision riding — and that of its Viennese counterpart, the Lipizzaners of the Spanish Riding School — the ballet left me cold. I remember sitting in the dark in that oper-alike building as a single beam of light caught a gray horse between pillars, performing the piaffe to music, first under the instruction of his trainer and then as the trainer walked away. The display went on for a long, long time, and I wondered later what might have convinced that horse to stop. The same horse galloping in a pasture would have given me more plea-sure than that ostentatious show of equine obedience.

In actual battle, such cool gymnastics as the pesade often fell prey to the rider's own fear and adrenaline or proved to be of little practical use — once foot soldiers had seen one pesade they had seen them all. And as the English discovered at Poitiers in 1356 and again at Agincourt in 1415, the long bow could take away any advantage in horse and knights. Further, when soldiers formed a four-sided "infantry square" and did not panic and run as their forebears had done, they could indeed repel a cavalry charge. By the 1830s the *haute école* style of riding moved from the battlefield to the circus.

First, of course, the circus — or at least something more permanent and elaborate than the dancing bears of the Middle Ages — had to be invented. That task fell to Sergeant Major Philip Astley, a British cavalryman. He opened a riding school in an old London lumberyard, but when enrolment flagged he put on exhibitions to entice recruits. He did somersaults and handstands on horseback. He would ride around the ring standing with one foot on the back of one horse and the other foot on another horse; the two horses would then jump over a pole, with Astley still on board. Crowds grew and Astley added clowns and jugglers and slack-wire artists, inspiring his competitors to became bolder: in 1772 David Wildman made a name for himself by galloping with one foot on the saddle and one foot on the horse's head while his own head was en-veloped in bees. Astley built his first fully enclosed amphithe-ater in 1780, then added musicians, tightrope walkers, clowns, jugglers and performing dogs. The Ringling Brothers would one day thank him.

Meanwhile, over in the New World, the horse still represented the paragon of military might. Until the advent of the repeating rifle and the revolver, Indian warriors on horseback completely outmatched the pioneers. It took a minute in the early days to measure and pour powder, ram the ball down the barrel, prime the tube, adjust the cap or flint and then fire. In that time, the Indian warrior could advance three hundred yards and let fly with up to twenty arrows.

Some historians believe the Comanche were the finest horsemen of the West. To "ride like a Comanche" became a western proverb. If horses were the mark of wealth among Plains Indians, the Comanche were the wealthiest of all. Each warrior might own fifty to two hundred head, and one chief, A Great Fall by Tripping, was said to own fifteen hundred. The numbers of horses owned by tribes defies the imagination: the great camp of Sioux and Cheyenne gathered along the Little Big Horn in 1876 was said to include from twenty thousand to forty thousand horses.

With astonishing speed (the horse came to them only after 1700), the Comanche exploited the horse in a military way. Boys of five were soon capable of saddling their own ponies and riding off with bow and arrow. Colonel Richard Dodge, an American cavalry officer, called the Plains Indian boy of twelve to fifteen years of age "the best rough rider and natural horseman in the world."

An American army captain once offered a large sum of money for the favorite horse of Sanaco, a Comanche chief. He declined to sell his buffalo horse lest his people suffer. Besides, Sanaco said, stopping to pat the pony on the neck, "I love him very much."

Proud warriors painted their horses as well as themselves. War horses had their tails braided with bright ribbons and feathers, their manes adorned with scalps. The horses wore necklaces of bear claws, with human hands or horse tracks painted on their flanks to indicate scalps taken or horses stolen.

There is no consensus on which tribe became the finest horse people. The Crow, the Sioux, the Comanche, the Blackfoot might all claim the honor, but the claim is beyond verification. Because they were among the first to see and use the horse, the Apache also warrant consideration. The Apache chief Geronimo (whose rifle I have seen in Monty Roberts's house, its stock notched in a kind of inverted Braille to spell *men*, each dot a recorded kill), rode a "blaze-faced, white-stockinged dun horse," according to his biographer. In close combat, a warrior was sometimes forced to leave his horse and escape on foot. Geronimo had trained his horse to come when called, and this the chief would do when he had found a safe place.

By the time of the American Civil War, the horse was employed more to move guns than to move men. Still, at least in the beginning, more horses and mules were killed than soldiers and the life expectancy of a horse at the front was six months. Some died of starvation, wounds or disease; some went mad from the noise and carnage and had to be shot.

Cavalry had a role to play in the First World War, but much less of one in the Second World War. Remembered because rare, the cavalry charge had by this time become a romantic gesture. Russian cavalrymen on tough little Siberian ponies did, however, successfully attack German tanks immobilized by frigid cold. And a bold plan to drop — by glider — almost twelve hundred mules and 250 horses into Japanese-held Burma turned out a great success for the Allies. The maneuver kept supply lines open for beleaguered battalions.

Horses were also used in the 1950s in Kenya against Mau Mau rebels. And during the Korean War a mare named Reckless was given the rank of sergeant for her valor in hauling ammunition.

In certain parts of America, the cavalry tradition lives on, with mounted RCMP officers patrolling the parking lots of malls. RCMP, in this case, stands for Royal Courtesy Mounted Patrol. It was launched in 1988 by a young horseman named

Frank Keller, whose company, Alpha and Omega Services, now has riders patrolling malls in half a dozen states. As Elizabeth Atwood Lawrence found in her field research, muggers and car thieves are extremely wary of horses, and the clip-clop of hooves actually makes people in neighborhoods feel warm and secure.

Keller's first hurdle in launching the RCMP was to establish a set of standard operating procedures that would satisfy his insurance company. He found what he needed by consulting old U.S. Cavalry field manuals in the Library of Congress. The horses must pass extraordinary stress tests that include exposure to firecrackers, squirt guns, smoke, helicopters and the wail of fire trucks. The troopers, who wear blue uniforms with gold trim and sit astride elegantly groomed horses, are unarmed. The horse is deterrent enough. One mall saw its crime rate fall from thirty-five incidents in one December to twelve the next.

The genuine RCMP, ironically, rarely use horses today: the famous "musical ride" and ceremonial occasions mark the only time a mountie sits in the saddle. The RCMP dress uniform — scarlet tunic and brown ranger hat — has become a Canadian icon recognized around the world. But as their name implies, the Royal Canadian Mounted Police rode to fame on the back of a horse.

The government formed the force in 1873 to bring order to the vast Canadian West, which was likely more lawless (and the American West more lawful) than legend would have it. The massacre that year of at least thirty Assiniboine Indians, victims of Montana wolf hunters on the trail of horse thieves, hurried the move. The so-called Cypress Hills Massacre in what is now Saskatchewan marked a turning point in the taming of the West. The task of the North-West Mounted Police, as they were then called, was extraordinary: three hundred men on horses were to patrol an area of three hundred thousand square miles. That they pulled it off speaks of their boldness and, on more than one occasion, their good fortune.

Imagine this scene from 1877. Sitting Bull and five hundred of his warriors, fleeing U.S. Cavalry in the wake of the Little Big Horn, are gathered at Wood Mountain in what is now southern Saskatchewan, where Sioux numbers have swollen to four thousand. The NWMP's task is to keep a lid on the unruly Sioux, who love to steal horses — even police horses. An NWMP officer, with a small escort, rides into Sitting Bull's camp and demands the return of the police horses. The chief — himself on a fine horse — almost laughs at him.

"I would take even the horse you are riding if I thought it stolen," says the officer.

"It is," replies Sitting Bull, drawing a line in the sand with his words.

What happens next seems cinematic (the Mountie legend did, after all, inspire some 250 films), but it's true. The officer edges his horse close by the Sioux chief's, yanks him out of the saddle and takes the horse by the bridle. Other officers close ranks around their commanding officer, then all gallop off to the nearby fort. For whatever reason, no retaliation ensues.

Other stories contributed to the Mounties' reputation for always getting their man (that motto, by the way, is pure Hollywood invention). At the turn of the century, Sergeant J. C. W. Biggs trailed a horse thief from Moose Jaw, Saskatchewan, into Montana in a marathon pursuit that spanned twenty-seven hundred miles and 135 days.

The sleek black mounts ridden today by the RCMP in their mesmerizing cavalry drills look nothing like the scruffy ponies initially used in the territory. Mountie history is replete with stories of unbroken horses throwing their first riders to the ground, but come January and the blizzards, those same horses, if their riders trusted them, brought them safely home.

I admire the instinct to honor the war horse, though not always the honor itself. I am puzzled, for example, by the strange custom of military men who pay tribute to great war horses by stuffing them or mounting their bones in glass cases.

To be fair, it is not just men in uniform who cannot bear to bury the equine dead. Roy Rogers, the film and TV cowboy from the 1950s, rode his palomino, Trigger, until the inevitable day came and the stallion stopped rearing forever. Or so it seemed. Roy "just couldn't put the old fellow in the ground with the worms and everything," so he put him to work, instead — with the tourists. Trigger deserved better than what he got: being stuffed in the Roy Rogers Museum in California, frozen in his famous hooves-pawing-the-air pose.

No ashes to ashes, dust to dust for Arkle, the brilliant Irish steeplechaser, either. His bones are on display in a museum at Kildare. As for Phar Lap, the legendary speed horse who died mysteriously in America in 1932, his heart and skin went to Australia, where he raced to fame, and his skeleton to New Zealand, where he was born. Even the rodeo world has gotten into the act: the rodeo bronc War Paint lives on, taxidermically, in Pendleton, Oregon.

But soldiers do beat all when it comes to twisting horse remains into horse monuments. General Philip H. Sheridan rode a horse called Winchester in the Civil War. So grateful for his services was the general that when Winchester died Sheridan had him done up like a big-game trophy and presented to a war museum. He (the horse, not the general) is still on display — at the Smithsonian Institution in Washington.

Another American general, Robert E. Lee, fought for the South and rode a storied gray called Traveller. The mere sight of the general on that horse was enough to inspire Confederate troops. "As he rode majestically in front of my line of battle, with uncovered head and mounted on Old Traveller," wrote a fellow general, "Lee looked a very god of war." The horse would outlive his master and join his funeral procession. In 1871, Traveller himself was buried, but not for long.

In 1907, the army disinterred his skeleton and parked it in a museum, where Traveller remained on duty until 1962, when he was finally put in the ground just outside the chapel at Washington and Lee University in Lexington, Kentucky.

Since Lee was buried in the chapel, horse and rider remain, you could say, close.

Little Sorrel, Stonewall Jackson's charger in the Civil War, also warranted a decent burial but instead ended up on display in the Soldier's Home in Richmond, Virginia. Stuffed horses, however, fell out of fashion, and in 1997, Little Sorrel's bones were cremated and his ashes sprinkled under Jackson's statue at the Virginia Military Institute. Tossed into the grave, as well, was earth from fourteen battlefields where the horse had served, along with a few carrots and some horseshoes.

Stuffing a horse seems strange to us now. But a century ago it was a token of an officer's affection for a trusted mount who may have saved his life many times over. In the land of the stuffed horse, one ranks above the rest. He is Comanche, the lone survivor (on the U.S. Cavalry side) of Custer's Last Stand, the Battle of the Little Big Horn, in what was then called the Montana Territory. We know a great deal about Comanche, thanks in large part to Elizabeth Atwood Lawrence. Her book on Comanche, *His Very Silence Speaks*, goes far beyond biography and amounts to a thoughtful treatise on the horse as both warrior and living symbol.

Comanche was a good-looking dark buckskin horse ridden by an Irish-born officer in the U.S. Seventh Cavalry, Myles W. Keogh. The horse's original name is unknown, but Captain Keogh called him Comanche, either because the horse was once wounded by the Comanche in 1868 or because the horse let out "a Comanche yell" when he took that arrow in the hip. Comanche dutifully brought Keogh back to camp, where the horse patiently and stoically allowed a farrier to pull out the arrow and dress the wound.

There would be many more arrows and bullets for Comanche, the foolhardy George Armstrong Custer and the horse soldiers and horses under his command. For the Sioux and Cheyenne warriors massed and waiting for Custer and his troops, the battle that June 25, 1876, was a resounding victory, the last of its kind. For the American people, it was a shattering defeat. When troops arrived in the Valley of the Little Big Horn two days later, they found every one of the 276 soldiers dead and every horse killed or gone — except Comanche. Amid the gore, he was the only living creature.

Little Soldier, a seventeen-year-old Sioux who fought at the Little Big Horn, later explained why that one horse was left. Captain Keogh, it seems, had ordered his men to shoot their horses and use their bodies as cover. He himself had not yet done so and maybe could not bear to. Keogh had a reputation as a loner, a romantic and a melancholy sort. He was also a gifted rider. A cavalryman who served under him during the Civil War said he "rode a horse like a Centaur."

But the centaur was surrounded. He was kneeling between Comanche's legs and shooting at his attackers when he finally died of bullet wounds. Keogh still held Comanche's reins tightly in his hand, and the Sioux looking on believed it was dangerous to disturb such a potent connection between rider and horse, between the living and the dead. "No Indian," said Little Soldier, "would take that horse when a dead man was holding the rein."

Every trooper at the Little Big Horn was stripped, scalped and mutilated, except for General Custer (respect for leaders was a tradition among Plains Indians) and Captain Keogh. Strangely, Custer also died with the reins of his horse, a fine sorrel, attached to one wrist. A Santee Indian named Walks-Under-the-Ground ignored superstition and took the horse that belonged to "Long Hair."

By the time the cavalry reached Comanche days later, his six bullet wounds had exacted a near mortal toll and one soldier was tempted to end the horse's misery by cutting his throat. But two other men put water in their hats and urged Comanche to drink. Slowly he recovered and was eventually nursed back to health.

No one ever rode him again. Colonel Samuel D. Sturgis of the U.S. Seventh Cavalry, who lost a son at the Little Big Horn, issued special orders concerning Comanche that Lawrence

5.4   Comanche: he survived Custer's Last Stand to become an outpost pet
      and later a museum attraction.

calls unique in the annals of military history. General Orders
No. 7, which included a phrase that gave Lawrence the title of
her book, read: "The horse known as 'Comanche' being the
only living representative of the bloody tragedy of the Little
Big Horn, Montana, June 25, 1876, his kind treatment and
comfort should be a matter of special pride and solicitude on
the part of the 7th Cavalry, to the end that his life may be pro-
longed to the utmost limit. Though wounded and scarred, his
very silence speaks more eloquently than words of the desperate struggle against overwhelming odds, of the hopeless conflict, and heroic manner in which all went down that day."

From that point, Comanche led the life of a military hero.
He was never again put to work, and it became a court-
martial offense to strike him or to ride him. Only on special
occasions would he be saddled and bridled and led by a
mounted trooper. He was housed in a handsome stall and

given several attendants, including a blacksmith named Gustave Korn, whom he adored. Comanche had become a living legend, and what Comanche wanted, Comanche got.

He acquired a taste for the finer things. During his long recovery, he was fed a whiskey bran mash every second day, and he would often wander over to the canteen, where the men offered him, and he willingly drank, buckets of beer. Much the outpost pet, he also solicited sugar from the men. Comanche would turn over garbage while foraging and sometimes would be seen with coffee grounds on his mouth — like a kid with a milk mustache.

No area was off-limits for this horse. His stall door always remained open so he could roll at will in the nearby mud wallow and graze by the bandstand during regimental concerts. He also helped himself to lawns and gardens, where he developed a fondness for the sunflowers painstakingly grown by officers' wives. Comanche would follow Korn around like a faithful puppy, and when the blacksmith went to a house to visit a lady friend, the jealous horse would neigh outside until Korn finally came out to lead him home. Comanche's caretakers claimed that when Korn passed away, the horse — by then a very old horse in any case — "lost interest in life" and died.

In 1891, they stuffed him. Many years passed before they put him in a glass case, and by that time thousands of visitors had come to the museum at the University of Kansas in Lawrence just to see the horse and touch him for good luck. Some could not resist taking a souvenir hair from Comanche's tail. One report has it that so many tail hairs were plucked the museum caretaker secretly replaced the tail seven times.

War horses often knew instinctively who was friend, who was foe. In a previous battle with Indians, after his men had slaughtered every one, Custer ordered the killing of eight hundred captured Indian ponies. This effectively destroyed the Indians' valued possessions, but was it not also a waste of good horseflesh? No, because Indian ponies clearly saw (or, more precisely, smelled) whites as the enemy and typically would not let soldiers even get close to them. Indian women, who could approach the horses easily, were forced to round up the doomed ponies. In the same way, U.S. Cavalry horses were just as afraid of Indians. Each side could, if desired, ride the other side's horses, but the rider needed both time and patience.

Countless anecdotes from military history point to the horse's heightened sensitivity to the presence of the enemy. A Canadian officer tells the story of a polo pony mare at a front-line veterinary hospital who could detect — long before soldiers could — the approach of aircraft during the Second World War. That alone is not unusual. But this mare could apparently distinguish between Allied and enemy bombers. If the planes were German, she would stop eating, throw up her head and, with ears erect, stand perfectly still. She would listen, then stamp, paw and show signs of excitement. But if the planes were friendly, she calmly continued eating.

Another Canadian soldier described a rather cunning mule — a gun-pack animal — who would find a hollow in the ground and lie low when she and her train came under fire from German machine guns. There she stayed until the firing ceased. But savvy alone does not explain the bond between soldier and horse, or why some war horses were held in such high esteem.

General Jack Seely (later Lord Mottistone) led Canadian mounted troops in the First World War and wrote a book, *My Horse, Warrior*, about the fierce love he felt for his war horse. A line drawing at the front of the book faithfully captures, says Seely, Warrior's "white star and his fearless eye." The horse had presence: when Seely rode among the men he did not hear "Here Comes the General" but "Here's Old Warrior."

The book pays tribute to the horse's great courage under fire and his indomitable spirit. If Warrior became the general's "passport" and won the affection of soldiers, explained Seely, it owed only a little to the horse's looks, ancestry and quality. Warrior impressed everyone who met him with his character and personality because he still possessed, despite the horrors of war, an air of innocence. He had never been beaten.

"The soul of a horse," wrote the general, "is a great and loyal soul, quite unspoiled by the chances and changes of human kind. Above all, it is a courageous soul, and an affectionate soul. But let there be one cruel blow from a grown-up man, and you have ruined the horse's fine soul and spirit for ever. It is my dream that those who read this book may vow never to beat a willing horse. Warrior has never been so beaten, partly by good fortune, partly because it takes a brave man to beat him."

After the war, Warrior retired to a paddock overlooking the English Channel. The old horse turned a benevolent eye to the children who came to stroke him. And whenever he saw Seely's wife, he would walk up to her, "lay his soft nose against her cheek, and close his eyes — the supreme tribute of friendship from a horse to a human being." The book ends with a little sketch of Seely in a tweed jacket riding off into the sunset on Warrior, who has his left eye turned slightly, looking back.

Like Seely, many soldiers developed a real affection for the horses they rode into war, sometimes over the course of years. In the American Civil War, for example, men under cover of darkness would often steal food for their horses; officers, knowing what feeling had prompted it, would turn a blind eye. And it seemed the affection went the other way, too.

General William B. Bate fought with the Second Tennessee at the Battle of Shiloh, in which he rode a magnificent black stallion called Black Hawk (formally called Canada Chief, owing to his Canadian pacer bloodline). After first being repulsed in battle, Bate told his troops to follow him, adding that he would not bid them go where he himself would not. Soon enough, Bate was wounded and fell off Black Hawk. The horse followed the men into battle for another half mile before turning back to locate his rider.

Black Hawk somehow — perhaps by tracking Bate's scent

or a trail of blood — followed the path of the horse-drawn ambulance carrying Bate to a hospital tent, which was three miles from the battlefield. The horse "poked his head in the tent door," Captain Robert D. Smith is quoted in the *Horse Review* of 1896, "and affectionately whinnied to his master while the surgeon was dressing the wound." The horse's own wounds, meanwhile, had escaped everyone's notice. Black Hawk walked a few paces into the woods, staggered and fell dead. Smith recorded how the horse's obvious attachment deeply moved Bate: "He can still see that almost human look Black Hawk gave him and that last pathetic whinny as he walked off to fall down and die."

One British cavalryman who penned a diary during a military campaign against the French in 1794 near Antwerp noted: "A soldier would as soon see his comrade killed as his horse; and that the horse has an equal regard for, and knowledge of, his master will be seen in the following fact . . ." The diary went on to describe a skirmish in which a dragoon was hit by a bullet and fell from his horse. Fellow soldiers, who were retreating from the swarming enemy, presumed he was dead. Two days later, a patrol discovered the dragoon's horse close by the body, and the pattern in the grazed grass clearly showed that the horse had never ventured more than a few yards from the downed rider. "When they buried the corpse on the spot," the diarist wrote, "the faithful animal seemed to show great reluctance to come away without his master, frequently turning his head and neighing, as if wishing his dead master to come and mount him; this was an old horse that had lived with one rider many years."

A former U.S. Cavalry officer told of seeing his war horse, Pig, years after leaving the army. The horse was now part of a six-horse team, and although the animal appeared miserable his former master could not afford to buy him back. "I went up to him and petted him," the old soldier wrote. "He knew me alright. He nickered and looked at me as much as to say, 'Come on, please, Charlie, get me out of here.' I had ridden old

Pig thousands of miles and more than once he saved my life. I pretty near cried when I saw him that time in the Black Hills."

In 1995, a film called *In Pursuit of Honor* illustrated — as only Hollywood can — the bond that sometimes formed between cavalrymen and horses. Apparently based on a true story, the movie stars Don Johnson as a tough American cavalryman in 1932 who joins others in a wild adventure to save doomed horses.

Convinced that the horse was obsolete as a military force, the army had ordered 506 cavalry horses at a base in Texas dispensed with, but not by selling them or even by giving them away. They were to be taken to a gully across the border in Mexico, machine-gunned and the earth bulldozed over them. "There's nothing left," says the Johnson character in a drunken rage. "No horses, no cavalry, no honor."

The cavalrymen, bonded with those particular horses and horses in general, watch the first lot of one hundred "murdered" and then risk court martial and death to save the rest. With the army in mechanized pursuit, five men, driving the herd before them, make a run from Texas to the Canadian border.

No doubt some cavalrymen felt they owed a debt to horses they had ridden for many years and who, like Pig, had saved their lives more than once. During the Second World War, a Red Army cavalry division was ordered to advance over a snow-covered minefield. The divisional commander told his men to ride with loose reins and let the horses choose their own path. Most of those who did so made it through. Most of those who rode collected were blown up.

There is an innocence about horses, a general willingness to please, that makes what happens to them in war nothing short of heartbreaking. A poster from the First World War shows a man holding in his arms the head of a dying horse. All around is the blasted landscape of that war — a shelled house, branchless trees, tortured metal on the roadside. Another soldier is urging the man to come. Up ahead an artillery wagon

"GOOD-BYE OLD MAN"

5.5   Some soldiers formed powerful bonds with their horses and grieved
when their comrades died.

pulled by two other horses is struggling in the mud. But the man, who is cradling the horse's head as he would a child's, must first bid farewell. "Good-bye old man," the caption reads. And the poster pleads, "Help the Horse to Save the Soldier. Please Join the American Red Star Animal Relief, National Headquarters, Albany, N.Y."

Animal relief. There is almost none of that in war. In the First World War, some 1.5 million horses were used as cavalry alone; an estimated five hundred thousand died. The diaries of cavalry officers paint a bleak picture of muddy battlefields where mange (a skin disease) and glanders (a swelling of the neck glands) spread among horses like wildfire. To control mange, the army had all horses clipped, but then they shivered in the cold. Water troughs froze; food supplies dwindled. Ravenous horses chewed and choked on halters and blankets, the manes and tails of other horses, their own hay nets, even the epaulets on men's shoulders.

Many thousands of horses caught what soldiers called "moon blindness," an eye disease linked with unsanitary conditions. Parasites were endemic. Mustard gas would leave the horses in terrible pain and, three weeks later, blind. Chlorine gas caused pneumonia and gangrene.

In his diaries, Captain Alfred Savage of the Canadian Army Veterinary Corps barely suppresses his rage over the abuse of horses during the terrible winter of 1917 in northern France. Four inches of snow lay on the ground and bone-chilling winds racked men and horses. More horses died in one week in January than during the previous six months, and two-thirds of the watering holes had been frozen for ten days. Since none of the officers using the remaining watering holes "seemed to give a damn," Savage wrote, "it was about time 99% of [them] received a course in Elementary Horse Management."

Later that year, at Passchendaele in Belgium, the earth was so rutted with shellfire that the ground acted like a sponge. Horses sank to their knees in the thick brown porridge. It was a mercy to shoot them. Lieutenant Colonel D. S. Tamblyn, a Canadian who fought there, wrote:

> The battle of Passchendaele Ridge was one of the terrible experiences of the Great War for our dumb friends. The return of pack animals and their leaders from the forward areas was a sight never to be forgotten. Horses and men, plastered with mud from head to foot, some exhibiting evidence of having received first aid treatment, others bearing ghastly wounds, and the carcasses of dead animals being used as stepping stones by the men to bring their charges and themselves out of the mire to more solid ground, made an awful picture. Shrapnel and bombs were bursting here, there and everywhere; horses neighing, and men bidding their last farewell to their dumb pals, who probably had been their chums since the commencement of the war, made matters worse and more unbearable. Such sights cannot but be indelibly impressed on one's mind.

With few exceptions, horses sent to war in Europe did not come home. Some died in transit. Across Canada sixteen horses were put to a boxcar, one thousand to a trainload, before they boarded ships at Levis, Quebec, or Halifax, Nova Scotia. More would die in the crossing; many succumbed to pneumonia from lack of shelter in England; countless more would die on the battlefield. At war's end, survivors were sold, often to butchers in war-ravaged countries desperate for food. Even in death, the horses served.

One of the last great battles involving soldiers on horseback occurred almost two centuries ago: the Battle of Waterloo in 1815. Two horses from that epic battle are still remembered.

That Bonaparte had twenty horses shot out from under him in battle underlines what a target the emperor must have presented on his white steed, and how vulnerable both were. The horse most closely associated with the conqueror, the horse depicted in most of those triumphant paintings, is Marengo, a name taken from a village in northern Italy where Napoleon and the French engaged the Austrian army in 1800. Napoleon's troops fought three separate battles that day. They lost the first two but won the final, decisive, one.

In that battle Napoleon rode a short, thickset gray stallion brought into France from Egypt only the year before. Just over fourteen hands tall (a short horse for a short man), the horse remained steady amid the chaos and the barrage of gunfire and cannonade, even though Napoleon himself was slightly wounded in the foot by a bullet that removed part of his riding boot. So pleased was the emperor with his horse that he named him after the battle. Marengo would take eight wounds over the course of his long military career.

The story is told that one day near the battlefield of Austerlitz in Austria, Napoleon was alone, deep in thought, as he led Marengo through the woods on foot. Suddenly, the horse tensed. Marengo snorted and pointed his ears forward. Napoleon leaped into the saddle and off they galloped, pursued by Russian spies hiding in the bush.

Marengo joined Napoleon in the folly of the Russian campaign and the subsequent retreat. By the Battle of Waterloo, where Marengo was wounded, he was a senior horse of twenty-two years. It was his last battle.

Napoleon used three horses, including Marengo, at Waterloo. In his hasty retreat to Paris after the defeat, he left Marengo behind at a stable near the battlefield. The victorious English took the horse back home, where he was royally treated and stood at stud for a long time, outliving his master by ten years. Marengo died in 1831 at the grand old age of thirty-eight.

By the way, they stuffed him, too.

Across the battlefield at Waterloo that day was a fifteen-hand chestnut originally trained as a racehorse and ridden by the duke of Wellington, also known as the Iron Duke. That horse, Copenhagen, similarly took his name from a battle.

Copenhagen had earned a reputation for stamina and

spirit. The day before the battle, Wellington rode him for ten hours and then from dawn to dusk at Waterloo. When the exhausted Wellington finally dismounted, Copenhagen swung around and aimed a kick that almost killed the duke. Next day came more mischief: the fiery horse eluded his grooms and had to be chased through the streets of Brussels.

British soldiers loved him. Painters and sculptors set down his likeness. And so much was written about him in diaries and memoirs that he approached the fame of his master. Lady Frances Shelley was granted a visit to Wellington's estate and, even better, a ride on Copenhagen. "A charming ride of two hours," she later wrote. "But I found Copenhagen the most difficult horse to sit of any I have ridden. If the Duke had not been there, I should have been frightened." When it came time, the duke gave Copenhagen — the subject of an obituary in the *Times* of London — the kind of funeral reserved for high-ranking officers. The duke did not, as far as I know, stuff him or mount him in a glass case.

I do find the notion of mounting the mount, as it were, distasteful. But even the stuffed horse, it seems, offers a service. In the last years of his life, Comanche became a powerful symbol for a nation in mourning. Elizabeth Atwood Lawrence maintains that horses can become the focus of widespread grief and hope. In *His Very Silence Speaks*, she cites the intriguing case of Sefton, a black gelding who survived an Irish Republican Army (IRA) bomb blast.

Sefton was part of the Queen's Household Cavalry and was in London's Hyde Park — no doubt posing for tourists' cameras — on July 20, 1982, when the first bomb went off. Four troopers and seven horses were killed, along with seven infantrymen (members of a military band) in a separate explosion in Regents Park.

Of all the surviving horses, Sefton was the most grievously hurt. The nineteen-year-old horse took thirty-eight shrapnel wounds and would have died of a severed jugular vein had a soldier not put his fist into the horse's neck to staunch the flow of blood.

An Irish journalist later remarked on the curious response to the twin bombings. It was the crime committed against the horses, as much as or more than the one against the humans, that stirred the greatest outrage in both England and Ireland. That week, wreaths were laid where the horses died; the dead bandsmen were not remembered that way. Medical updates on the surviving horses earned greater prominence on television newscasts than did those of the wounded men.

Like Comanche, Sefton would henceforth be no ordinary, and certainly no anonymous, horse. A painting of "the brave and defiant horse" was auctioned off to raise money for victims' families; the painting, in turn, led to cards, plates and pendants. A great wave of get-well cards urged the horse to live and, by doing so, to "cheat the bastards [the IRA]." When the horse did indeed survive and appeared in a public ceremony for the first time, out came the hankies. "The sobbing," said one observer, "was contagious and quite unashamed."

What to make of all this? Is the emotion not misplaced? Lawrence, an anthropologist, sees it in a more positive light. "By diverting attention from grief," she wrote in *His Very Silence Speaks*, "as well as by its own presence, an animal possesses a remarkable capacity to heal." Lawrence points out that Custer's horse Dandy (not present at the Little Big Horn and therefore spared) helped assuage the grief of Emanuel Custer, who lost three sons, a son-in-law and a nephew in that battle. "I don't know how I could have lived without that horse," the elder Custer once wrote. "He's been a comfort to me for thirteen long years." Elizabeth Custer called Dandy her father-in-law's "anchor."

Comanche played the same role for General Sturgis, who lost a son at the Little Big Horn, and for the American nation, which had suffered a crushing defeat. All that grief was transferred onto one horse.

British author and ex-cavalryman J. N. P. Watson, who wrote a book on Sefton, believes that the horse's leap into the limelight has everything to do with the innocence that almost all equines possess. The horse, he said, is so lacking in malice and yet so dutiful and brave that "when he suffers it makes man ashamed for the human race."

The silence immediately following the bombing unnerved witnesses. Not one horse, even Sefton with his ghastly wounds, snorted or whinnied. One eyewitness called the quiet "so absolute it chilled the mind and the senses."

Like Comanche, Sefton enjoyed a glorious retirement. He had always been a horse comfortable at the center of things. A British brigadier recalled feeding him mints and patting him, this while the horse was recovering from surgery. And when the man turned to talk to someone else, Sefton gently but firmly grabbed his hand and brought it back. "He wanted my attention again," said the brigadier, "and knew how to get it. He has a great personality, a very strong personality, and that was an important factor in his recovery."

At the Home of Rest for Horses, it was forbidden that Sefton be ridden. Crowds came around with carrots and apples and mints (eventually the treats were rationed, lest he balloon or colic), and no one seemed particularly surprised that Elizabeth Atwood Lawrence had traveled all the way from America to visit this celebrated horse. Sefton died, finally, in 1993 at the age of thirty, whereupon they put him where he belonged. In the ground.

CHAPTER 6

THE WONDER HORSES
OF HOLLYWOOD
AND LITERATURE

Tonto: "Him shine like Silver."
Lone Ranger: "Silver! That would be a great name for him."

Roy Rogers: "We were getting ready to do the first picture.
I was fooling around with my guns as we talked. I believe
it was actually Smiley Burnette who said, "As fast and as
quick as the horse is, you ought to call him Trigger, you
know, quick-on-the-trigger. I said, 'That's a good name.'"

I HAVE A LITTLE scar at the base of my neck where the gun butt struck. My memory of the incident is kaleidoscopic, with blank spaces and faces and one last name. That little crescent moon where the hair never again grew is as real as the images: the blood trickling down my back, the trip

6.1   Wee cowboy, 1927: a fascination with the cowboy way led in turn to
a fascination with horses.

home cradled in linked arms, my rescuers all talking at once as the door opens, the look on my mother's face . . .

My assailant, Ernie Cichelly (to end this little tease), was seven. Home then was Nakina, an outpost in northern Ontario. It was a spit of a town, with an odd name that matched others on the railway line: Capreol, Hornepayne, Gogama. I remember my father once crossing the tracks and taking me into the bush, where the bugs left scabs in my scalp the size of dimes; how he made me an Indian bow, with arrows to match. Later, again by the tracks, I proudly handed the bow to an older boy who had expressed interest in it; he snapped it over his knee before passing it back with a laugh. I remember tears, but mostly shock that the world was capable of such malevolence.

Although suddenly bowless, I could still be a cowboy. Minimal attire was a cowboy hat with drawstring, a bandanna at the neck and a metal six-shooter in a leather holster. (*Serious* quick draws wore one on each hip.) At the end of the holster the leather split off into two thongs that you could wrap round your leg. As we had seen them do in the movies, we would "knock out" the other guy from behind with the handles of our cap guns. There was an art to delivering a menacing-looking blow but stopping the gun butt just shy of the other kid's head; Ernie had yet to master it.

Later, as I grew up in suburban Toronto in the late 1950s, the emulation of cowboys continued. We supped on westerns at the Golden Mile Theatre and on television. Our first set was a windowed piece of furniture that also housed the hi-fi. At first, we ignored the thing. The giant cardboard box it came in was far more interesting as a fort, a cavalry's defense against imaginary tribes.

Decades later, this period in my life would become a rich source of trivia questions:

Q. Who played Hopalong Cassidy?

A. William Boyd. (The black-attired Hopalong and his gray horse, Topper, made the cover of *Time* magazine on

November 27, 1950, the year their names appeared on 108 products, from wristwatches to kids' clothes, thermoses to ice cream. The endorsement bonanza was worth $70 million.) You got bonus points if you knew that my childhood hero at first detested children ("the little bastards") and feared horses.

Q. Who said, "Wait for me, Wild Bill"?

A. Jingles, played by Andy Devine in the *Wild Bill Hickok* TV series, uttered those words in the opening credits while jiggling atop a bay horse.

Q. Name the horse.

A. Joker.

Sometimes I wonder if my own affection for horses quietly took root in those early days of television. The Cisco Kid rode Diablo; his sidekick, Pancho, rode Lobo; Wild Bill Hickok rode Buckshot. Five horses named Silver, Scout, Trigger, Buttermilk and Champion mattered as much as the five characters who rode them on late-afternoon TV: the Lone Ranger, Tonto, Roy Rogers, Dale Evans and Gene Autry. The gray, the two paints, the palomino and the Tennessee walking horse entered our hearts as horses in books had before them — Flicka, Black Beauty, National Velvet.

The twentieth century, on the one hand, took work from the horse with every passing decade. The war horse, the plow horse and the ranch horse all fell from view, but in their stead came a far grander version of the horse to fire our imaginations and keep the horse-human bond alive: the heroic horse of the silver screen and the TV screen.

While cowboy extras earned a pittance for their dangerous work, star horses made small fortunes for their trainers. Fury, the black American saddle horse stallion who starred in a TV series in the 1950s, worked only four months a year and still pulled in $500,000 in eight years. Cowboy film stars, meanwhile, also grew rich. Gene Autry earned close to $100,000 in 1937, and by 1948 he owned his own movie and music publishing companies, oil wells, cattle ranches and a traveling rodeo, bringing his *net* income that year to $600,000.

6.2   In 1949, children were buying two million Roy Rogers and Trigger
comic books every month.

Trigger starred in all eighty-seven Roy Rogers films and
101 TV shows, before finally succumbing in 1965 at the age
of thirty-three. Like Topper, Trigger was a merchandiser's
dream. Some sixty-five different Roy Rogers–Trigger products
flew off the shelves in 1949, and their comic books were selling at the dizzying rate of two million a month.

On rainy-day Sunday afternoons I would sometimes watch

on TV, one of the seven *Francis* the talking mule movies made between 1950 and 1956. For these, the trainer tied thread to the mule's lips or, off camera, pressed a muscle on the side of the animal's face to create the illusion of equine gab. Television emulated this, and Mr. Ed, a palomino, talked his way through his own show for three years in the early 1960s. There was no sense, least of all horse sense, in any of this; there was, though, continuity. Alan "Rocky" Lane, a cowboy star of the 1940s and 1950s and a solid horseman, seemed a natural to give voice to Mr. Ed. And the same director oversaw both Donald O'Connor in the silly *Francis* movies and equine Ed in the even sillier television series.

While children watched television shows about horses by day, in the evening during the late 1950s and 1960s adults watched *Bonanza, Gunsmoke, Cheyenne, Bat Masterson, Maverick, Have Gun Will Travel, Wanted: Dead or Alive, Death Valley Days, Iron Horse, Tales of Wells Fargo, Laramie, Wyatt Earp, Rawhide, The Virginian* and *Wagon Train. Gunsmoke* (1955 to 1975) and *Bonanza* (1959 to 1973) rank among the longest running episodic programs in the history of television.

Television and film exalted the horse, with plots almost as improbable as those of the silent films. The Lone Ranger — silver bullets, mask and all — had rescued his magnificent wild gray after an attack by a buffalo, nursed him back to health and was repaid a hundred times over by the thankful stallion.

Many cowboy stars (John Wayne, Tom Mix, Gene Autry, Hoot Gibson, Clark Gable and Sal Mineo, to name a few) shared the silver screen with eternally grateful mustangs. As nonsensical as some were, such films may well have helped Wild Horse Annie and her consorts save the real mustangs, who were being slaughtered even as their celluloid counterparts were being celebrated.

It did seem to matter that the horse ridden by film and television characters was not just a horse but a horse with a *name*: Tex Ritter rode White Flash; James Arness in *Gunsmoke* rode Buck; in *Wyatt Earp*, Hugh O'Brien rode Candy.

6.3   Lee Marvin on Smoky in the film *Cat Ballou*: the memorable morning-
after shot.

By the 1960s horses no longer occupied center stage in
western films but still were cast in strong supporting roles. In
Cat Ballou, Lee Marvin plays the dipsomaniacal Kid Sheleen,
and in a morning-after scene, the Kid is on his horse, Smoky,
sleeping it off in the saddle and leaning against a saloon wall.
The horse hilariously mirrors his rider: with his legs crossed
and his head down, Smoky looks almost as hung over as the

Kid. When Lee Marvin won an Academy Award in 1966 for his performance, he acknowledged in his acceptance speech Smoky's role: "I think half of this," he said, referring to the Oscar, "belongs to a horse somewhere out in the Valley."

By this time, the presence of the horse was likewise waning on television. Lorne Greene's Pa Cartwright rode a horse, all right, but I never learned his name. This was a far cry from the old days, when the horse's stardom (and acting skills) matched the cowboy's. The end of programs such as *Bonanza* and *Gunsmoke* marked the virtual end of the western as a staple of film and television. It had been a good long run, one that began with the old western horse operas. In fact, the very creation of moving pictures in the late 1800s owes something to the horse.

The ancient Egyptians and every subsequent culture that relied on the horse debated this question: is there a moment in the horse's trot or canter when all four feet are off the ground?

The man who finally settled the matter did it in 1872 with photographs taken in quick sequence. His name was Eadweard Muybridge, and he had been commissioned by Leland Stanford, former governor of California, a railway mogul and owner of a breeding ranch, to determine through photography whether Stanford's horse Occident ever had all four feet off the ground during his splendid trot. Stanford had bet a friend $25,000 that Occident did.

At Stanford's ranch in Sacramento, California (now the site of Stanford University), Muybridge arrayed a battery of cameras in rows and triggered the shutters electromagnetically. His accomplishment was a stroke of genius, because rolled film and automatic shutters had yet to be invented: like old rifles, old cameras took only single shots. Muybridge connected clocks and circuits and cameras and obtained exposures as short as $1/6,000$ of a second. As Occident passed the shutters went off in perfect sequence, like a neat row of dominoes falling in quick succession.

6.4   Photographs like these, taken by Eadweard Muybridge in 1887, led to
motion pictures and a starring role for horses.

The resulting photographs clearly showed two points in
each stride where Occident *did* leave the ground. Stanford won
his bet. (It was, though, a Pyrrhic victory, since Muybridge
racked up $100,000 in expenses over five years to produce
irrefutable evidence.) If scientists were intrigued, so were
artists. For centuries, painters and sculptors had gotten the
horse's gallop, trot, even the walk, wrong. Drawings of galloping cattle done by the Bushmen of the Kalahari Desert
show that they, at least, understood four-legged locomotion.

Muybridge's horse photographs were first published as
The Horse in Motion and later joined other photos collected
in *Animals in Motion*. The tiny black-and-white images run
across the page, comic book-style. If you scan left to right,
you can watch a man in a bowler galloping a horse named
Daisy, her mane flying, and in two sequential shots all four
legs are in the air. On one page, fifteen frames capture a near-

naked man jumping a fence on horseback — on the approach, in the air, on landing. Cats and dogs, deer and birds, naked women and horses, lots and lots of horses, swiftly cross the pages in the first stop-action photography the world had ever seen. Muybridge suggested that his device could be used to determine the winner of horse races. And so it was: the photo finish was first used in 1888.

Muybridge also invented a "zoöpraxiscope," a device to project images sequentially on a screen. He was soon chatting with another inventor, Thomas Edison, about creating *talking* pictures by marrying the zoöpraxiscope and the phonograph. Someone did just that.

Inspired by Muybridge, Etienne Jules Marey in 1887 invented the chronophotographic camera, which could record the motion of a trotting horse on a roll of light-sensitized film. This machine led directly to the motion picture camera.

Seven years later, Edison made his first film, one of the very first films ever made, a 643-frame short called *Bucking Bronco*. He followed that in 1896 with *The Burning Stable* and its sequel, *Fighting the Fire*. Both vignettes depicted similar scenes involving four white horses being led from a fiery stable and horse-drawn fire engines arriving to extinguish the flames. Other Edison horse shorts, *Elopement on Horseback* and *Cripple Creek Bar-Room*, were made in 1898.

The silent film had arrived. Not long after came talking films, then talking films in color. The horse would cling to his starring role in this new medium for at least fifty years. Until well into the 1950s, horse and cowboy were the focus of thousands of cheaply made movies. They called them "B movies," and for a long time I was convinced the B stood for bad. They were actually movies meant to be the second feature on a double bill, in the days when a single cinema ticket let you see two movies. Plot-poor but action-rich, "B movies" featured runaway stagecoaches, burning wagons, posse chases, warriors on fleet paint horses and cavalry to the rescue.

In these stylized and predictable morality plays, the horse

was king, a real character who rescued the hero in the white hat and outwitted the bad guys in the black hats. Some movie horses were as famous as the actors, and even received more fan mail. The horses' pictures and names took pride of place on posters outside theaters.

Fritz, a red-and-white pinto in films of the 1920s, was perhaps the first horse to get a credit line in a movie. He was ridden by one of the great actors in the early westerns, William S. Hart, who thought Fritz deserved equal billing.

Unlike some actors in later years (who could neither act nor ride), Hart was a capable actor and a fine horseman. He built whole films around Fritz and once wrote a book in "the voice" of Fritz. "I loved every hair of the little scoundrel's hide," Hart said in his autobiography. The horse was precious, both to Hart and to his fans.

With his small but extraordinary stunt horse, Hart did the "run and throw," where the horse in full gallop suddenly stopped and his rider threw him to the ground. They jumped through windows and over fire and crossed raging rivers. One film in 1924, *Singer Jim McKee*, shows Hart riding his horse over a cliff before a treacherous 150-foot slide down a gorge. Movie fans bitterly complained about cruelty to their beloved Fritz, but they need not have worried.

Hart, who was said to love Fritz more than he ever loved another human, thought the stunt too dangerous even for his gifted horse. The filmmakers therefore constructed an elaborate dummy horse animated by piano wire. The footage obviously fooled many people. Hart placated his fans and the censors by traveling to New York, where he showed clips to explicate the fake horse. But the issue of Hollywood's cruelty to horses was destined to come up again.

In those days the mark of stardom was to sink your hands in wet cement outside Grauman's Chinese Theatre in Hollywood. Tom Mix, a famous cowboy of this era, had a horse called Tony who would eventually have his hoofprints grace that sidewalk.

Tarzan, yet another wonder horse, starred in films of the 1920s and 1930s. His partner was a cowboy hero named Ken Maynard. The author Edgar Rice Burroughs, creator of the jungle hero, had apparently visited the Maynard ranch and suggested the Tarzan name. However, an ensuing lawsuit over its use suggests that Burroughs either did no such thing or had second thoughts.

A $50 horse with palomino looks, Tarzan was astonishingly versatile and learned to respond to one-syllable commands. He could dance, bow, roll over and feign death; he could nod in response to questions, ring a fire bell, untie the ropes that bound his master and jump from great heights. Maynard's character was forever getting stuck in quicksand, finding himself surrounded by hostile Indians, or being knocked unconscious in burning buildings. Tarzan always came to the rescue, and he would nudge the reticent hero into the arms of the heroine for the kiss that ended the movie.

In his book *The Filming of the West*, Jon Tuska called Tarzan "the most exceptional horse in pictures. . . Ken loved Tarzan deeply, in a manner that he never loved a human being. Tarzan reciprocated."

Tuska tells the story of the two attempting a stunt in a film called *Gun Gospel*. Horse and rider were to leap from a cliff sixty feet down into a lake, but Tarzan slipped on the planed runway (put there, ironically, for safety's sake) and somersaulted before landing in the water on top of Maynard and driving him thirty feet below the surface. Each swam for opposite shores — Tarzan for the far shore, as he had been trained; Maynard for the near shore after almost drowning. When he spotted his rider, Tarzan immediately swam back to Maynard, who waded out to meet him. There, in the shallows, the horse dropped to his knees and nestled his head in Maynard's arms.

Another horse called Dice (Gregory Peck rode him in a film called *Duel in the Sun*) could pull a revolver from a pocket, lift cowboys by the seat of their pants and smile or yawn on command. In one picture Dice, a pinto stallion with a mostly white face, walked through a hotel lobby, entered an elevator, then backed out and began to climb the stairs.

I doubt that audiences believed in the cleverness of these celluloid horses, any more than they swallowed the saccharine truths of the Shirley Temple films, but moviegoers did delight in the fantasy of horses-to-the-rescue. Two films produced during the First World War were called, respectively, *Saved by Her Horse* and *Saved by His Horse*. In one film called *Trail Through the Hills*, released in 1912, Indians tossed the hero off a cliff, and the cowboy's pony later showed up at the precipice, dropped a rope and pulled up his two-legged pal. (A logical sequence might have seen horse and rider retire to the saloon for a chat and a beer.) In another film, we were to believe that the horse galloped in front of an outlaw and intentionally took the bullet meant for his master.

H. F. Hintz, the author of an exhaustive survey called *Horses in the Movies*, observed that the notion of a horse, a flight animal, entering a burning barn to untie his master's hands "was nonsense. But it was nonsense we enjoyed."

Some of the actors in these films — Tom Mix, Ken Maynard, Hoot Gibson — had been rodeo stars or stuntmen. The fine Canadian writer Guy Vanderhaeghe wrote a novel in 1996 entitled *The Englishman's Boy*, which touches on the lives of cowboys hanging around Hollywood movie sets in the 1920s, hoping to be hired as extras, stuntmen and doubles. The author describes how compounds were built to house the sullen men until a director came along to cut one or two "out of the remuda for a day's shooting."

In those days, lead actors often performed the stunts themselves. But they also had doubles, and so did their horses. Actors in the next generation, such as William Boyd, were sometimes poor or reluctant horsemen. Gene Autry, a green rider, got only marginally better. Jack Palance, who played the mean boss on the cattle drive in the *City Slicker* movies with

Billy Crystal and acted in countless westerns before that, did not at first ride tall in the saddle.

In that classic western *Shane*, released in 1953, gunslinger Palance *walks* his horse into town, and the effect is both chilling and dramatic. In fact Palance walked the horse because the director's instruction to gallop him, even a subsequent plea that Palance trot him, was then beyond the actor's equestrian skills.

The word on John Wayne was that cowboys on the movie sets liked him — the horses didn't. Corky Randall, a second-generation Hollywood horse trainer, recalls Wayne as "a rough rider. He never squeezed with his legs to make a horse go, just kicked him with his spurs." After two movies, a replacement horse would have to be found because the original horse had learned to anticipate spurring and would launch himself into a full gallop when the word "Action!" was uttered. Clint Eastwood, on the other hand, says Randall, "knows how to sit on a horse."

But Clint or Duke sitting on a horse was nowhere near as interesting as Clint or Duke in a gallop. Hollywood wanted action — warriors on paint horses circling a wagon train where John Wayne, say, is seen aiming his Winchester; cut to the next frame, where a warrior takes a bullet in the chest and his pony goes down violently in a cruel rain of dust.

Horses of the 1920s and 1930s who became famous by being partnered with prominent cowboy actors were prized and often highly trained, some even to fall on command; and what they could not do, their doubles did. But the equine extras, the "wild mustangs," the run-of-the-mill ponies ridden en masse in posses and Indian bands — how were they made to fall with such dramatic effect? Tripwires were often used, and the injuries and deaths that occurred among animal actors were simply part of the price to be paid for entertainment. "Horse spectacle," the movie directors labeled it.

The tripwire device was called "the Running W." It had its roots in ranch work when cowboys would cure a horse of running away by tying a rope from the horse's front legs, through a ring on the saddle and up to the pommel. A fleeing horse soon fell, but just before he did the cowboy would yell "Whoa." The horse learned to avoid a fall by obeying the call. Yakima Canutt, who won an Oscar in 1966 for a lifetime of Hollywood stunt work, first deployed the Running W to produce dramatic shots of horses pitching to the ground.

It worked like this. Off camera the film crew drove a post, called a "deadman," deep into the ground. Two lines of piano wire were strung from just above the horse's hooves up the front legs and then back to the girth. Beneath the horse's belly, these lines met inside a closed metal band shaped somewhat like a W (thus its name). A single line of wire was led out from behind the horse for several hundred feet, coiled beside the deadman and the end securely fixed to the post.

The stuntman's job was to ride a horse at a hard gallop to a spot, sometimes marked by a kerchief in the sagebrush, where the line would all be played out. At that point the wire, invisible to the camera, stopped the horse cold, yanked out his legs and sent him crashing to the ground. Ejected from the saddle, the rider hoped to spare his bones by rolling. The horse, of course, got no warning. Sheepskin-lined hobbles at the fetlocks, affixed with rings, spared the horse cuts to the legs, but that was all he was spared. Yakima Canutt claimed to have done three hundred Running W leaps without ever crippling a horse. He blamed amateurs for the fact that Running Ws killed countless horses and hurt or maimed many men.

The Running W demanded extraordinary expertise on the stunt director's part if death or injury was to be avoided. The piano wire had to be precisely gauged so it broke without too much strain as the horse's front legs were drawn up to his chest. Otherwise a complete somersault that broke the horse's neck or back was likely. Even if the horse was spared physical injury, the psychological damage from the severe and unexpected fall — the horse's loss of confidence — was permanent.

In *The Hollywood Posse*, the author Diana Serra Cary recounts her father's life as a Hollywood double but also offers the horse's perspective. Early in the book a horse the director calls the "stunt bay" is introduced to the new stuntman: "Jack checked the cinch on the big bay, who was edgy as hell. He fought back when the prop man threaded the wire through the leather hobble on each foot. It was obvious the bay had been put through this stunt before, and didn't take a shine to it. His muzzle and the white blaze on his face were crossed with scars where the hair had grown back a different color."

In this scene Jack is supposed to be a member of a posse chasing outlaws, one of whom turns in the saddle and shoots Jack dead. Jack must go down hard. Which he does, along with the big bay. Imagine Jack in the lead with riders alongside, all in a full gallop. Jack scans the ground for the flash of red bandanna that will give him his warning. As soon as he sees it he kicks both feet from the stirrups, and instantly the ground rises up to meet him. He keeps rolling amid the dust and "a forest of horses' legs" and is miraculously not trampled by the bay or the posse. When horse and man stand, each considers the other.

The dazed horse seems astonished to find himself alive after yet another felling, and the stuntman feels he owes the horse an apology for putting him through it. "That's one helluva way to treat a good horse," he says as he leads the bay, piano wire in tow, back to the rope picket line.

The Hollywood Posse includes some distressing photographs of horses who did *not* find themselves alive after one of these action shots. One photo shows a horse on the ground, the hobbles still on the fetlocks, the piano wire clearly visible. He has been shot by someone from the American Humane Association after breaking his leg in a manufactured fall. The cutline reads: "He was dead. No more work on location that day. The general attitude was of men ashamed of their jobs. Among the spectators, a woman wept bitterly. She knew the horse well, for it belonged to her husband."

Another photo, taken during the filming of *The Three Mesquiteers* in 1940, shows four horses going down at once in Running W scenes: two horses have struck the ground hard with their noses. Some of these films were made in five days on low budgets; doubles and horses were expendable items. Cary argues that *The Charge of the Light Brigade*, shot in 1936, features some of the most dangerous group Running W stunts in movie history. Close to a thousand horses were used on the set of that film, which saw horses dying every day during shooting. Stuntmen were also injured daily and one died when he broke his neck in a fall.

The ruthlessness of directors became a cliché. In a scene in *The Englishman's Boy*, Guy Vanderhaeghe depicts a slow-witted stuntman named Miles on a big black gelding named Locomotive. The horse is what the cowboys called "a croppy" — his cropped ears posted a warning of his mean temper. The stuntmen all refuse to ride Locomotive, despite the director's taunts and challenges to their manhood. Promised a part in a future film, Miles finally steps forward.

Shorty, a cowboy at the center of this mesmerizing novel, does what he can to help poor Miles. He paces out the length of piano wire and pegs a marker into the ground. "Miles," he says, "when you see that there white hankerchief coming up on you, you kick your feet out of them stirrups because when that fucking widow-maker runs out of wire he'll go ass over tea-kettle and when he does you ain't going to want to get hung up in them stirrups — you going to want to get throwed clear. Throwed clear, understand? Otherwise, you going to smash up bad, like an apple crate."

Miles spurs Locomotive into a full gallop and is watching for the handkerchief, but suddenly the horse is somersaulting with Miles's feet still planted in the stirrups. Locomotive lives, as does Miles, but the rider limps badly and passes bloody stools for the rest of his days. There *was* no handkerchief to be seen: the director had instructed a camerman to remove it surreptitiously. The all-important action shot was worth more than a man's, and certainly a horse's, life.

In the early version of *Ben-Hur*, filmed in 1925, up to 150 horses were reportedly killed and several men severely injured. Posed publicity shots from that film depict clearly dead horses in a chariot pileup. Much like a Roman emperor, the producer had offered rewards ($150, $100 and $50) for the first three drivers over the finish line. But the 1959 version, starring Charlton Heston, apparently occasioned no serious injuries to man or horse despite many crashes and spills. The directors had taken enough care, or so they claimed, that no animals died on the set. Something had happened between *Ben-Hur*s to change the way Hollywood used animals in general and horses in particular.

The turning point had come in 1949 with the filming of *Jesse James*, shot at Lake of the Ozarks in Missouri. The film gained infamy among animal rights advocates.

It starred Henry Fonda and Randolph Scott, and near the end the brother train robbers are fleeing a posse that has them trapped at a precipice. One brother whacks his horse on the flank. In the next scene, shot from the lake about seventy-five feet below, the horse tumbles head-over-hoof into the water before landing in a vile spray of water. Then the film cuts back to the cliff, where a second horse gets slapped and is seen tumbling. Accounts of the episode vary: we may be seeing the second horse falling or, more likely, the first unfortunate animal — same leap, but this time shot in a tighter frame, with more light and from a different angle. In the ensuing frame, the two brothers are swimming alongside their horses and the posse is left behind.

But the scene at the cliff did not fool everyone. One, possibly two, horses, died that day. One version in a book on the history of stuntmen reported that the horses originally shied in terror at the drop, so chutes had been built and the stuntman and two horses pushed off the cliff. Outraged viewers complained loudly and bitterly. That same year the American Humane Association (AHA) was founded to govern the use of animals in film.

The result today is that any film employing animals typically includes an explanatory note at the film's end assuring viewers that no animal was harmed during the shoot. "We even protect insects," says Ed Lish, a field training officer with the AHA in Los Angeles. "Anything in front of the camera, you cannot harm. We look at scripts beforehand, we mark the animal action and make recommendations on how to do this or that safely."

Doubtless, abuses still occur, but not on the scale that once was the custom. Falling horses are now trained to do just that, and then only onto prepared beds of sand. Horses working around guns and explosions are fitted with cotton in their ears. And the tripwire has been banned. When a horse leaps from a cliff, the height is trimmed to ten feet or less, and only trained horses are given the task.

I did not lack for horses as a child. Film and television teemed with them. At my grandfather's farm near Tamworth, Ontario, where we often spent summers (compensation for Nakina winters), I had them in the flesh — giant plow horses, a gray named Queen, a bay called Molly. My grandfather saw the farm dog as a working animal, ignored the cats, left the hens and geese to my grandmother, and while I can remember him calling gently to the cows (co-bas, co-bas), it was the horses he prized. They were the only animals Leonard Flynn really talked to.

I have a photograph — black-and-white, of course — of me at four years of age standing on a hay wagon, reins in hand. I look stern and my gaze is forward — to the horses. Maybe I absorbed some of my grandfather's passion. But at some point, when I could read, literary horses came into view.

The first horse book I ever read was Anna Sewell's *Black Beauty* — a Christmas present. It was hardcover, with each of its glossy color plates lying under its own little blanket of snow-white onionskin paper. That book, along with Daniel Defoe's *Robinson Crusoe* and Robert Louis Stevenson's *Kidnapped*, would occupy a special place in my little library. It is psychologically telling, I suppose, that they are nineteenth- and eighteenth-century books, respectively, about horses, solitude and voyage. The boy who read them would grow up a Luddite who prizes horses, periods of quiet time alone and travel.

I liked *Black Beauty*'s simple elegance, rejoiced in its sense of comeuppance. When Dick the plowboy tosses stones at Beauty and the other colts, he is sacked for his crime by the farmer. Another boy stealing Beauty's oats is similarly caught. But Beauty's life and fortune are tied to his masters, this one kind, that one ruthless. "I hope you will fall into good hands," Beauty's mother tells him. "But a horse never knows who may buy him, or who may drive him. It is all a chance."

Surely no coincidence, many writers who first led us into the inner lives of horses were women. The contemporary writers of horses I most admire are women: Maxine Kumin, Elizabeth Atwood Lawrence, Vicki Hearne. The old stereotype is still more true than not. Girls and women tend to befriend their horses and to rely on gentle persuasion. Boys and men are more likely to use muscle in an attempt to dominate the horse. Women writers were drawn to the tender territory of horse and child, and they plumbed it long before male writers did.

Marguerite Henry wrote at least ten books about horses, including *Misty of Chincoteague*. Mary O'Hara created *My Friend Flicka* and others in that series. But it was Anna Sewell's *Black Beauty* that set the tone for the others to follow. Suddenly, beast of burden and warrior horse had to make room for noble companion and equine personality.

We live in an age of numbers, so perhaps some numbers about Sewell and her book are in order. I chanced upon a Penguin edition of the book published in 1972, which looked much thumbed and yellow with age. Eleanor Graham had written the introduction and noted there how difficult it had been to find a publisher for the book — a first book written by someone past fifty and on her deathbed.

When *Black Beauty* was published on November 24, 1877, London bookshops ordered fewer than one hundred copies for the entire city. But by 1935 the publisher had notified Sewell's niece that world sales were in the vicinity of twenty million copies. By the 1970s the book was thought to be the sixth-best seller in the English language.

It had taken Anna Sewell seven years to write — proof once more that time, not just talent, creates fine and enduring books. Moral vision does, too.

Anna Sewell was a deeply religious woman, born a Quaker, who used Quaker "thees" and "thous" all her life. After an injury, she was left lame in both feet and navigated outside in a pony carriage. She would hold the reins loosely in her hand, never using the whip and always talking to the pony as if the pony understood.

Sewell had read an essay on animals by Horace Bushnell, who argued that animals, as much as humankind, live to do the will of God. That notion haunted her. "I have for six years," read a note found amid her private papers, "been confined to the house and to my sofa, and have from time to time, as I was able, been writing what I think will turn out a little book, its special aim being to induce kindness, sympathy, and an understanding treatment of horses."

The Sewell family believed that Black Beauty, the horse at the center of the book, was modeled on Bessie, Anna's brother's horse. They recognized Bessie's spirit and courage, good sense and affection, and her uncommon speed.

Black Beauty chronicles the life of a horse and is told from that horse's perspective, giving voice to every equine in the book. Here is Beauty on the sensitivity of a horse's mouth: "Oh! if people knew what a comfort to horses a light hand is, and how it keeps a good mouth and a good temper, they surely would not chuck, and drag, and pull at the rein as they often do. Our mouths are so tender, that where they have not been spoiled or hardened with bad or ignorant treatment, they feel the slightest movement of the driver's hand, and we know in an instant what is required of us."

In one episode in the book, a clear sample of the horse's acute sensibility, Black Beauty is about to take his driver,

John, and his master, Squire Gordon, across a wooden bridge during a violent storm in near darkness. Beauty stops, refusing to cross, even after a light touch from the whip. At that moment, a man at the tollgate runs out with a torch to warn them that the bridge was washed out in the middle. Had they gone on, all would have perished.

In an ensuing conversation Beauty hears his master say that "God had given men reason, by which they could find out things for themselves; but He had given animals knowledge which did not depend on reason, much more prompt and perfect in its way, and by which they had often saved the lives of men." The incident at the bridge may well have been based on historical precedent. I have heard and read similar stories.

Anna Sewell would also have known the fate of horses in war. The Charge of the Light Brigade, that folly inflicted on men and horses only a few decades before *Black Beauty* was written, must have been fresh in her mind. The accounts I have read seem mirrored in Sewell's book as Captain, an old war horse, tells his story to Beauty:

> From the right, from the left, and from the front, shot and shell poured in upon us. Many a brave man went down, many a horse fell . . . Fearful as it was, no one stopped, no one turned back . . . Some of the horses had been so badly wounded that they could scarcely move from the loss of blood; other noble creatures were trying on three legs to drag themselves along, and others were struggling to rise on their fore feet, when their hind legs had been shattered by shot. Their groans were piteous to hear, and the beseeching look in their eyes as those who escaped passed by and left them to their fate, I shall never forget.

Marguerite Henry surely read of Black Beauty. She wrote several classics of her own: *Misty of Chincoteague*, *King of the Wind* and *Black Gold*. There is enough subtlety in the characters and enough genuine detail to counter Henry's own insistence that the heart be warmed, that the tale end happily.

A horse called Black Gold did indeed win the 1924 Kentucky Derby. The book gives a keen sense of that horse, his line and the people close to him. In his youth, the horse's Irish jockey, Jaydee Mooney, had tended the horses that pull funeral coaches. One of his jobs was to "take the vinegar" out of the horses by riding them just before the funeral so they would look more stately in the procession. The job taught him something important about riding: "The chief thing, he discovered, was to be one with the horse, to be part of him, motion for motion."

Black Gold is out of a filly named Useeit (she was so tiny as a foal she could barely see out the half door of her stall, and thus the name) and by a fine Kentucky Thoroughbred stallion named Black Toney. It is Useeit, an Oklahoma filly bred by Osage Indians, almost as much as her famous foal, Black Gold, who wins the reader's heart.

The tale is clear, never muddied; the writing, admirably simple; the metaphors, aptly chosen. As Useeit matures, she got "round and solid as an apple. And her eyes, always beautiful, became so full of health and liquid light that one was stopped by their brilliance." Her brown coat had a sheen, "like a plain brown boulder made glossy by the water that flows over it."

In real life, the horse's trainer told his wife on his deathbed that he wanted Useeit bred to Black Toney because he had a vision of the mare throwing a Derby winner.

In true storybook fashion, Black Gold is born with his father's great endurance and his mother's breathtaking speed. But his owner, a selfish and misguided old man, fails to take action after the Derby win when the horse develops a crack in a front hoof. Black Gold, meanwhile, still loves to race. The bugle call from a nearby private track sets him off, and every day he conducts a little race in his paddock, wearing a circle on the perimeter.

Finally, during an actual race in New Orleans, Black Gold

snaps a leg above the ankle; only tape holds it in place. "But," as the track announcer puts it, "he finished his race — on three legs and a heart he finished it." He was buried at that track, Fair Grounds Park, where every year the winning jockey in the Black Gold Stakes puts a wreath of flowers on the grave.

Human affection for horses may owe something to generations of readers returning over and over to the well of literature on horses. *My Friend Flicka*, a book that dates from 1941, begins: "High up on the long hill they called the Saddle Back, behind the ranch and the county road, the boy sat his horse, facing east, his eyes dazzled by the rising sun."

The boy is Ken McLaughlin. He is desperate for a colt of his own and he has spotted a young mustang filly. He calls her Flicka (it means "little girl" in Swedish) after a Scandinavian ranch hand referred to her that way. A book such as this might be wrapped up today in eighty pages in one of those thin-as-gruel horsy paperbacks aimed at young readers. But *My Friend Flicka*, a 251-page paperback, has endured for close to sixty years.

The somewhat churlish Rob McLaughlin, the boy's father, is a compelling character. On the one hand, he kills cougars, harbors no romantic sentiments about feral mustangs and wants to shoot Flicka when an infection brings her close to death. On the other hand, he rails against traditional horse breaking: "I hate the method, waiting until a horse is full grown, all his habits formed, and then a battle to the death, and the horse marked with fear and distrust, his disposition damaged — he'll never have confidence in a man again."

More out of efficiency than softheartedness, he wants the horses treated well. The rancher lectures his son Howard for being heavy-handed, telling him that while horses sometimes need to be punished they should never get more than is necessary. He praises Ken for his delicate hands, but laments his dreamy ways that often let the horse take control. He tells both sons to be in the corral when a certain horse is to be broken by a rider with fine hands, a light seat and perfect balance: their mother, Nell.

The way she approaches the trembling, fearful mare calls to mind the modern gentler's approach. Nell is slow and patient, lets the horse smell her and then, significantly, turns her back to the horse to talk to the men. "Under the eye of a human being," Mary O'Hara writes, "an unbroken horse is in terror." When the horse nuzzles her back, Nell carefully turns and begins talking to the animal. She strokes her, leans on the saddle and places her knee under the mare's belly as if to mount, and only when the horse shows no signs of fear does she rise into the saddle and begin the work.

The strange thing is this: Mary O'Hara wrote *My Friend Flicka* when she was fifty-six and only after she had moved to Wyoming with her second husband. A strong woman who had made it as a scriptwriter in Hollywood in an era when most women felt confined by domestic duty, she somehow sensed that a three-thousand-acre sheep ranch thirty miles outside Cheyenne, Wyoming, was where she was meant to be. In her autobiography, *Flicka's Friend*, she concedes she could never have written the book without actually living in Wyoming and seeing wild horses. Wyoming, she came to realize, was not another state but another planet, one that "orbited at a different tempo, under different skies, under different orders."

Wyoming made her, as it did me, "silent and spellbound."

O'Hara, meanwhile, struggled to write and sell stories to magazines. One day she realized that whenever she chattered on about her animals, people stopped and listened. And whenever a story was written about wild horses in the American West, a long line formed at the Cheyenne library to read it. What if, she wondered, she were to write a tale about that mustang filly she had once seen caught in barbed wire? What if, in a story, a little boy were to catch and tame that young horse? Within twenty-four hours, O'Hara had the notes and scenes that would form *My Friend Flicka*.

Some years later, O'Hara was taking a course in short-story writing at Columbia University in New York, an older woman embarrassed to be in a classroom of young, aspiring writers. (As with Anna Sewell, her first book would come late in life.) One evening the instructor announced he would read three manuscripts, the last being the Flicka tale. The whole class fell on it, as if upon a feast in the forest. The instructor put O'Hara on to an agent called Sidney Lambert, who later told O'Hara: "I've been saying it was time for another *Black Beauty*. You've written it."

Lambert wisely advised O'Hara not to cash the $25 check sent her by the writing instructor when he published the story in his anthology. Had she signed that check — and the little contractual hook on the back — and not kept it as a souvenir, the anthology's publisher would have been entitled to half of all future revenues. Given that more than four million copies of the book have been sold to date, those revenues are substantial. Flicka had a future: the "vulture" of a writing instructor knew it instinctively, and so did Mary O'Hara.

Proof that Flicka lives on came during my week of riding in Wyoming in 1997. One of my fellow riders was a fifty-something woman named Ro (short for Rosemary) from Suffolk, England. She was arrestingly polite and formal, and her very Britishness seemed at odds with her cowboy hat and chaps. Her impish smile, however, served to warn of some small, impending mischief and I learned to heed it. I also came to realize that her passion for horses was rooted in knowledge and experience gained from raising Arabs at home.

As Ro would later explain in a letter, she had read the Flicka trilogy as a child and still escapes into those books. It was precisely because of those books that coming to Wyoming had seemed so much like a homecoming. "I spent most of my childhood and teenage years obsessed by horses and ponies, riding them and reading about them. Heaven was galloping into the wind — it was like flying. My imagination galloped too — I was a cowboy, an Indian, a Pony Express rider. I loved reading — Westerns were my favourite — not for the stories (I can't remember a single one) but for the horses."

Her zeal for horses existed in any case; the literature on horses had honed it, sharpened it, given it a language. For other people, too, reading that literature prepared the way so that when they came west, sometimes decades later, it felt familiar, even welcoming. On that Wyoming trip, I encountered several women, from Holland, Boston and New York, who had moved west and set down roots, sometimes in dramatic fashion. One woman had taken as a holiday a riding trip — a gift, ironically, from her husband; she fell for a horseman and that was the end of her Long Island life. The Dutch woman had similarly come west to ride for a week; she is now happily married to a wrangler. They slipped into the land of horses and never looked back.

Maybe it was like that for Will James, too. The award-winning author of *Smoky the Cow Horse* (published in 1927 and made into a Hollywood film in 1933) was born Joseph Ernest Nephtali Dufault to a francophone family in St-Nazare d'Acton, Quebec, in 1892. His parents ran a general store, and four-year-old Ernest would sit outside, transfixed by horses. His sister Eugénie remembers how he would draw horses without having to look at them: "He would just start with his pencil at the hoof and work his way up."

When he was nine or ten years old, Ernest mistakenly consumed a bottle of caustic lye and only alert action by his mother saved his life. Like many, many writers who have dealt with incapacitating childhood illness or injury by turning to books, Ernest spent his convalescence devouring cheap western novels left behind by guests at his parents' boarding house. The horse stories and romantic illustrations must have inflamed his cowboy yearnings, but what put him over the top was seeing the Buffalo Bill Wild West show in Montreal. His father later enrolled his wide-eyed son in art classes. Where other students sketched nudes, Ernest drew horses.

His mania for all things western finally led him to leave

6.5 Will James, left, with actor Randolph Scott in 1932: his love of horses was more genuine than his legend.

home, when he was barely fifteen, for farms and ranches in Saskatchewan and Alberta, and later all over the American Southwest. He adopted the dress, language and culture of the cowboy. Though he spoke no English until well into his teens, his joual gave way to a convincing drawl.

Dufault would refine both his drawing technique and his

writing style when he had some time on his hands — fourteen months in a Nevada prison for cattle rustling. He worked up a sketch for the character he would wholeheartedly become: Will James, born in a covered wagon in Montana; orphaned early when his father, a Texas-born cowhand, was lanced by a rogue steer; raised by a French Canadian trapper named Jean Beaupré (a clever fiction to account for young Will's accent).

Everyone bought the story. Adopting cowpoke grammar, but with a nice eye for detail and, especially, horse character, Will James the writer soon had his audience. In 1922, he sent a dozen sketches and a short story called "Bucking Horses and Bucking-Horse Riders" to Scribner's Sons in New York, where a legendary editor championed it. Maxwell Perkins, who had taken his red pencil to the novels of Ernest Hemingway and F. Scott Fitzgerald, would watch Will James become a legend himself.

Throughout the next two decades, James would write twenty novels (most with the words *horse*, *cow* or *cowboy* in the title) and a memoir called *Lone Cowboy*. The film of *Smoky the Cow Horse* begins with Will on camera offering this paean, delivered in true cowboy drawl, to the horse: "Well, folks, to my way of thinking, there's something wrong and amissing with any person who hasn't got a soft spot in their heart for animals of some kind. Me? I prefer horses 'cause they're man's most faithful and useful friend."

The voice, though, was dubbed. By then a binge drinker, James flubbed so many scenes and narrations that most were cut from the film. Even he had begun to believe the big lie he'd concocted in 1920; but the more he was pulled into the limelight, the more he feared being unmasked. He returned home and burned incriminating photographs and letters, yet his private terror still hounded him. He drank — to forget Ernest, to sink deeper into Will.

The fortune he amassed was soon squandered; the eight-thousand-acre ranch in Montana was taken by creditors. Will James died of cirrhosis of the liver and kidney failure in a

Hollywood hospital in 1942. Only in 1967 was the truth finally told: the American biographer Anthony Amaral discovered the James will, which had, either wittily or boozily, left part of the estate to Ernest Dufault in Ontario. Amaral followed the tracks to Auguste, Ernest's brother in Ottawa. The famous American cowboy's favorite meal, it turns out, was *tourtière*, a classic meat pie from Quebec cuisine.

What was genuine about James-Dufault, unlike his folksy spelling, was his love for the horse. "I admire every step that crethure makes. . ." he wrote in the preface to the first edition of *Smoky the Cow Horse*. "I've come to figger a big mistake was made when the horse was classed as an animal . . . He never whines when he's hungry or sore footed or tired, and he'll keep on a going for the human till he drops."

Like James's Smoky, other horses from literature have found their way into film. Anna Sewell's Black Beauty (half a dozen films and one TV series), Walter Farley's Black Stallion, John Steinbeck's Red Pony, Marguerite Henry's Misty — all have inspired Hollywood films many times over, all in celebration of the child-horse connection.

Behind the scenes, that connection brought at least one young actress to tears. In the film *National Velvet*, for example, produced in 1944, twelve-year-old Elizabeth Taylor played the role of a little Sussex girl who wins a horse in a raffle, disguises herself as a boy and rides the horse to victory in the Grand National Steeplechase. One scene called for tears as Velvet learns that her sick horse may not live.

Taylor's co-star, the veteran actor Mickey Rooney, counseled her that to produce real tears she might imagine great hardship and misery inflicted on her own family — her father dying, her brother starving. Elizabeth, who in fact felt no great affection for her family, had a better idea. "All I thought about," she said later, "was this horse being very sick, and that I was the little girl who owned him. And the tears came."

Taylor's genuine love for the horse imbued her character.

"I can't help it, Father," she says in the film. "I'd rather have that horse happy than go to heaven."

Taylor had ridden horses since the age of four. Although Monty Roberts was her double in some of the racing scenes, Taylor railed against using doubles and paid the price. She suffered a concussion during the filming of *National Velvet*, and in *Lassie Come Home*, filmed the year before, a horse stepped on her foot and broke several bones.

The horse, real or imagined, will always inspire writers, and anyone who reads widely will have encountered horses. Male writers have also found *Equus* a powerful subject: John Steinbeck and his red pony, Sir Walter Scott and Lochinvar, Sir Arthur Conan Doyle and Silver Blaze, D. H. Lawrence and St. Maŵr, Dick Turpin and Black Bess, Virgil and his warrior horse, Homer and Diomed, Leo Tolstoy and Frou Frou, Mark Twain and the Mexican Plug, Cervantes and Rozinante, Liam O'Flaherty and the old hunter, Rudyard Kipling and the Maltese Cat, Wallace Stegner and his colt.

Or those uniquely Swiftian horses in *Gulliver's Travels*. In 1723, at the age of fifty-six, Jonathan Swift (born in Ireland to English parents) undertook long journeys on horseback, including a two-hundred-mile trek along the wild coast of western Cork. Maybe he came to love horses then; maybe he always had. But in his famous satire, written in 1726, he seems to wish that men were more like horses.

In part IV of *Gulliver's Travels*, Swift writes of "A Voyage to the Country of the Houyhnhnms" (read, horses). The humanlike Yahoos on that island are nothing to admire; the horselike Houyhnhnms, on the other hand, were "so orderly and rational, so acute and judicious, that I at last concluded, they needs be Magicians." They are sweet and caring, and unashamed of their nakedness.

"The word *Houyhnhnm*, in their Tongue," wrote Swift, "signified a *Horse*, and in its Etymology, *the Perfection of Nature*."

Many of the men who wrote of horses earlier this century were eastern-born, even European-born, chroniclers of the cowboy way. Zane Grey, of Zanesville, Ohio, started his working life as a dentist, but a trip out west in 1904 changed all that and he began writing western novels, among them *Riders of the Purple Sage*. Fifty-four novels would net him more than seventeen million sales. Owen Wister, a Philadelphian who took a music degree at Harvard, intending to become a composer, earned his fame with *The Virginian*, a novel of Wyoming cowpokery written in 1902 and based in part on the life of an Alberta cowboy. That novel inspired the 1960s television series; similarly Larry McMurtry's *Lonesome Dove* inspired both a mini-series and a weekly television program in the 1990s.

The cowboy, not so much the horse, occupied center stage in these books and the authors played fast and loose with the truth. An American stagecoach guard once observed that "the majority of cowboys shoot pool better than pistols." But books and film entrenched the myth of the cowboy way, and it lives on.

The Indian way had its champions, too: some of them poseurs like Grey Owl (a Canadian conservationist named Archibald Belaney, who passed himself off as half Apache) and Karl May (pronounced "my"). May was a nineteenth-century scam artist in Germany who wrote novels featuring a German frontiersman, a warrior and a trapper named, respectively, Old Shatterhand, Winnetou and Canada Bill. Writing of locales he had never seen and getting lots wrong (he claimed all Indians could create a smokeless fire), he nevertheless inflamed German imaginations and would sell sixty-five million books, making him one of the best-selling authors in history.

Today, six hundred or so frontier clubs in Europe foster Western fantasies. Every spring some four hundred tepees are set up in a cow pasture in central Germany and hundreds of people dressed to the authentic nines — eagle feathers, bear claws, six guns and chaps — pretend they are cowboys and

Indians. *Der wilde Westen* lives on, thanks, in the main, to Karl May.

And reverence for the horse continues to mark literature. *Angela's Ashes*, winner of the Pulitzer Prize for literature in 1997, is a memoir by Frank McCourt, an Irish American writer who grew up penniless in Limerick in the 1930s and 1940s. His father is a hopeless romantic, Irish nationalist and drunk, and his mother has too many babies who die of hunger and deprivation. The family lives in a tenement house, the poorest of the poor. Nearby is a stable, where an old work-horse named Finn finally breaks down and has to be shot. Here is the scene through the eyes of eleven-year-old Frank as the horse is being drawn up a plank into a truck:

> The three men and the stable man tie ropes around Finn and pull him up the planks and the people in the lane yell at the men because of the nails and broken wood in the planks that catch at Finn and tear out bits of his hide and streak the planks with bright pink horse blood.
>
> Ye are destroyin' that horse.
>
> Can't ye have respect for the dead?
>
> Go easy with that poor horse.
>
> The stable man says, For the love o' Jaysus, what are ye squawkin' about? Tis only a dead horse.

Frank's mother and his brother, who had earlier stood vigil over the horse to keep the rats off the body, fly in rage at the man, who retreats into the truck. All this for a dead horse.

The play *Equus*, written by Peter Shaffer in 1977, is still produced, still shocks with its horse killings, beheadings and blinding. The horse is still sacred; harm to a horse, still heresy.

Something in the bond between horse and rider is so ancient, so archetypal and rich, that writers will never tire of exploring it. Redemption and seduction, danger and ecstasy, hope and despair — these are the territory of the horse. "How the horse dominated the mind of the early races, especially of the Mediterranean!" wrote D.H. Lawrence in *Apocalypse*. "You were a lord if you had a horse. Far back, far back in our dark soul the horse prances . . . The horse, the horse! The symbol of surging potency and power of movement, of action, in man."

In her contemporary story "What Shock Heard," Pam Houston sketches a cowboy remarkable for his quiet hands and calm. When a frantic horse called Shock balks at entering a trailer, the cowboy just whispers something in his ear; that and a carrot do the trick. A woman watching him ride sees, in effect, the centaur. Perfection: "I hung back and watched the way his body moved with the big quarter horse: brown skin stretched across muscle and horseflesh, black mane and sandy hair, breath and sweat and one dust cloud rose around them till there was no way to separate the rider from the ride." Later, the woman, not yet the cowboy's lover, asks him what he said to Shock to entice him into the trailer. The cowboy tells her there are no words for that.

CHAPTER 7

SPORT HORSE LEGENDS

*He is an athlete, a champion, and when you hold that
picture in your mind, or the picture of his closing run
in the Kentucky Derby, you have defined . . . a standard
for all the other horses you will ever see.*
PETER GZOWSKI ON NORTHERN DANCER,
IN *An Unbroken Line*

HORSE SPORTS ARE dangerous for horse and rider both.
Speed may come at a price; ground can be so unforgiving.

Dick Francis turned to writing novels — longhand, in pencil, one a year — when, at the age of thirty-six, his body could
no longer stand the rigors of the steeplechase (so called after
Irishmen named O'Callaghan and Blake agreed to a village-to-
village, church steeple-to-church steeple horse race in 1752).
The word *rigors*, though, politely skirts a shattering, orthope-
dic litany. Francis, a former champion, has fractured his skull,
wrist, jaw, arm and three vertebrae, broken his nose five times,

his collarbone a dozen times (twenty-one broken bones in all) and more ribs than he can remember. "You don't count broken ribs," he once said. "The pain stops when you warm up." Every night before bed the seventy-seven-year-old Francis must heavily bandage his left shoulder, the one he dislocated horribly in a fall from a horse.

Bones mend, though. Bodies heal. The hidden wounds, the ones of the mind, sometimes never do. In 1956, Dick Francis was riding in the Grand National at Aintree, a demanding course of thirty fences stretched over four and a half miles. I have never been to a steeplechase race, but footage invariably depicts horses crashing through hedges and riders landing ingloriously in heaps. The up-and-over motion reminds me of merry-go-rounds, especially because riderless horses continue leaping fences.

Francis was leading the pack that day on a horse called Devon Loch — owned by the Queen Mother — and galloping toward the finish line only thirty-five yards away. With the last fence behind them, the path to victory seemed clear. "In all my life," he would later write in his autobiography, *The Sport of Queens*, "I have never experienced a greater joy than the knowledge I was about to win the National." Never had he felt "such power in reserve, such confidence in my mount, such calm in my mind." He would have broken the speed record for the course. Would have, had Devon Loch taken ten more strides. No one, least of all Dick Francis, understands what happened next.

The horse's hind legs stiffened and he fell flat on his belly, spread-eagled cartoonishly. When the horse stood up he seemed paralyzed and not a little perplexed. A photograph records Francis tossing his whip in anger and dismay. He has his head buried in the horse's neck, as if he cannot bear to watch the pack streak past. Like the race car driver on empty a stone's throw from the finish line, he felt an anguish so great that even forty years later it remains a painful topic of conversation. "A post-mortem one day," Francis wrote, "may find the words 'Devon Loch' engraved on my heart, so ever-lasting an impression has that gallant animal made upon it." Did the crowd's roar unhinge the horse? Was it a medical problem missed by the vets — who later found nothing amiss despite exhaustive examination? Or was it a quirky shy?

The memory of the spills, the burden of being tagged the man who "lost" the National (a race he would never win) — these Dick Francis sets aside. What he chooses to recall is the adrenaline rush. One of his characters, a retired jockey in *Whip Hand*, awakens from a dream in which he has won a race: "I could still feel the irons round my feet, the calves of my legs gripping, the balance, the nearness to my head of the stretching brown neck, the blowing in my mouth, my hands on the reins."

A horse race, whether we partake or watch, awakens something in us. One galloping horse offers a vicarious thrill; a dozen horses on the fly fiercely pull us in. In the final moments of a race, who can turn a back to all those horse heads bobbing, those legs pounding the turf? Moderns feel the tug, as the ancients did.

The Romans were so keen on chariot racing that individual charioteers loomed as large in that society as Michael Jordan and Wayne Gretzky loom in ours. Roman fans knew the horses, their pedigrees, their quirks. On the floor of a Roman bath archeologists found a two-thousand-year-old tribute to a horse named Polydoxus: *Vincas, non vincas, te amamus Polydoxus*, the mosaic tiles read. "Whether you win or lose, we love you, Polydoxus."

The track, or hippodrome, drew huge crowds, and spectators wore blue or green to reflect allegiance to the charioteer of their choice. But the swelling numbers of Christian converts viewed horse racing as a pagan vice; invaders — Huns, Vandals, Goths — sacked Roman cities; earthquakes toppled hippodromes; and for a thousand years there was no racing.

Or at least no organized racing. But in fields and along dirt roads, challenges were surely laid down. Boys must have raced their colts; girls, their ponies; men and women, the pride of their stables.

The U.S. Cavalry and plains Indians engaged in savage fighting in the nineteenth century, but during lulls they often indulged in a favored pastime — racing horses. One day, at Fort Chadbourne in Texas, soldiers and warriors laid down their bets and the Comanche produced a most unlikely horse and jockey: an oversized rider on a pony with a three-inch-thick coat of hair. Throughout the four-hundred-yard race, the rider apparently used his weighty club on the pony and eked out a win — by a neck. The officers brought out a better horse, who also lost to the bedraggled pony — by a nose.

The cavalry had saved for the third and final race a Kentucky racing mare. But the Comanche bet everything they had on that miserable pony. When the gun went, the rider this time tossed his club, let out a whoop and took off. The trooper and the sleek Kentucky mare were soon so far behind that the Comanche rider took to sitting backward on his pony and beckoning them to catch up. The pony, it turned out, was a celebrated racer, and the Comanche had previously stripped another tribe of six hundred horses using the same sucker tactic.

The first Europeans to land on America's shores could not wait to race their horses. The otherwise dour Pilgrims were forced to pass laws against racing horses in the street. In the film *Friendly Persuasion*, Gary Cooper plays a Quaker tempted to race even on the way to church Sunday morning when a neighbor speeds by in his buggy. Not wanting to be seen succumbing to that temptation, yet keen to race, he cleverly trades for a horse with a minor vice; he will not abide being passed.

The first formal racetrack in America was built in 1665 in what is now Nassau County, Long Island. In *The History of Thoroughbred Racing in America*, William H. P. Robertson points out that while racing in England was the sport of kings, with courses designed by gentry for gentry, "The democratic

7.1 Eventer seeming to dance on water: to ride or watch an elite horse is
     to partake in that animal's grace.

spirit in America resisted such treatment. The public demanded that contestants in races should be visible from start

to finish, rather than disappear over a hill, and for this reason the circular track came into existence."

Canadians similarly longed to watch horse races and, more important, to bet on their outcome. Front Street in York (present-day Toronto) was the scene of legitimate street racing in the early 1800s. *The Plate*, Louis E. Cauz's history of the Queen's Plate, among the longest running stakes races in North America, describes racing's origins in what was then known as Upper Canada. "Trials of speed, one cavalry officer challenging another, or two farmers wagering livestock or produce to settle whose mare was fastest" had offered afternoon entertainment since the late 1700s.

The American Civil War would introduce new equine blood on both sides of the border. For safe keeping, many fine southern Thoroughbreds stayed in Canada and improved racing stock. The Canadian horse, meanwhile, a sturdy all-purpose roadster bred in Quebec, went south by the thousands.

The horse was on the move, and so was the horse sport. While the West produced the explosive Quarter Horse specifically for the quarter-mile dash, the East refined the Thoroughbred and Standardbred for longer races. Ranch work begat the rodeo, which pulled the cutting horse into competition. No other animal but the horse would be used in so many diverse athletic endeavors. No other sport but the horse sport would partner human and animal so directly or find so many ways for men and women to go head to head in competition.

Dressage, the hunt, the rodeo, polo, steeplechase, eventing, driving, show jumping — the theater of horse and human would make us, if not actors ourselves, then keen patrons, and sometimes both.

The literature of the sport horse is like a deep spring you return to again and again, so revitalizing is the water. The stories that touch me most are the ones that end suddenly and calamitously. In such a telling, there is no chance for the horse's gifts to erode, no time to second-guess the horse's worth as the career winds down and the losses mount. No, the fire burns hot and bright, then it's out.

I cherish the story of Ruffian, the stunning black filly. Walter Farley said she was the horse he envisioned (never mind her sex) as he wrote *The Black Stallion* books. One veterinarian called her the most perfectly conformed — best balanced, best proportioned — horse he had ever seen. But I remember the words of another vet, who took from the Devon Loch episode a lesson about the sublime: "The nearer to perfection, the more likely something has got to give."

A granddaughter, on the dam's side, of Native Dancer (a line that produced Northern Dancer), Ruffian was a fleet and spirited horse. She had a grace that belied her size and a quickness that staggered everyone.

Her long, fluid stride certainly confounded the exercise rider who first rode her on the track. The former jockey Yates Kennedy, then fifty-nine, had been around horses all his life, and Ruffian's ride seemed so effortless he thought he had gone the three-eighths of a mile distance in thirty-seven seconds. Like many other jockeys and show jumpers, he came equipped with a little stopwatch in his head. Rarely off, Kennedy was that day way off — by almost two full seconds. He later conjured an image to capture the sense of being on Ruffian's back. It felt as though she unfurled an invisible sail between strides, he said, so that when her feet were off the ground she rode the wind at her back.

Ruffian was born in April 1972 at Claiborne Farm, one of the grand old breeding farms of Kentucky. Colts rule Thoroughbred racing, but this filly was an exception. Some rank her not just the greatest filly who ever lived, but the greatest racehorse.

Her owners had already set the name Ruffian aside for a certain colt, but when he was sold they gave the name to the little filly. "Girls," insisted her owner, Barbara Janney, "can be Ruffians, too." She would be no delicate little mare. Stable hands nicknamed her Sophie, as in sofa. As big as a couch she

7.2 Ruffian: the Great Match Race of July 6, 1975, would be her last.

may have been, but she was not ungainly, as many large horses are. Quite the opposite.

Before trainer Frank Y. Whiteley Jr. put the jockey on Ruffian for her first training ride, he defied a superstition at the track: he sang her praises. Racing reveres luck — your starting position is drawn from a hat; a loss can be a matter of breaking a second too soon or too late, the mood of the horse, the footing, the weather. Wise track people therefore honor the tradition of withholding praise until duly earned. Best not rile the god of luck. But old Whiteley reckoned Ruffian's star would shine a long time, for the gods themselves had blessed her with special gifts. "I got a big black filly I'm gonna put you on," he told the rider. "It's the fastest horse you've ever been on."

In a short race against horses at the stable, the jockey,

Jacinto Vasquez, restrained Ruffian, but she breezed by the others. This despite a poor start and a trip around the outside to catch the leaders. The trick with Ruffian was not to find the gas pedal but the brakes. It took every ounce of Vasquez's considerable strength to stop her from continuing to run when the race ended. He was then a leading jockey on the circuit, thirty years old and notoriously tough. Vasquez's hands and arms went numb from pulling. Old Whiteley was right.

In her first race, on May 22, 1974, against other untried fillies, Ruffian opened up a fifteen-length lead and matched the track record. Ruffian won her second race by seven lengths; her third, by thirteen lengths. This would be her pattern. In nine of her ten races she either matched or broke the track record and only once was she pushed (that race marked the one time a crop was used on her, and only four cracks at that). In the tenth race, Ruffian was coming off an injury and the jockey was specifically instructed to coast to victory and *not* to break any records.

Racing magazines called her "a wonder" and said she was "invincible." "Speed to spare," reported the *Daily Racing Form*. "One was seeing something very rare," trumpeted the *Bloodhorse* scribe who admired her almost effortless acceleration. At Saratoga, her jockey tried to hold her back on the home stretch (he was that much ahead) and she still ran the six furlongs faster than any two-year-old in the history of this track where racing began after the Civil War. For all that, she had never really been tested and she had never run against a colt. How fast, went the whisper, could she run?

Someone proposed running Ruffian against the best colt of the day, the Kentucky Derby winner Foolish Pleasure. Just those two horses. The Great Match Race would be held at Belmont Park in New York on July 6, 1975. Boy versus Girl. Colt of the Year versus Filly of the Year. Bold Ruler's grandson versus Bold Ruler's granddaughter. The Race of the Century.

People wore buttons to the race, each one depicting a head shot of one of the horses, with the horse's name below and

"The Great Match" above. Some buttons simply read "Him" or "Her." Of the fifty thousand spectators at Belmont that day, many appeared to divide along gender lines. Men and boys wore Foolish Pleasure buttons. Women and girls demonstrated their support for Ruffian by wearing her button. One newspaper ran a cartoon showing women known for their strong views on women's rights — Gloria Steinem, Betty Friedan and Billie Jean King among them — all yelling "C'mon Ruffian!"

In most races, colts will outrun mares. Often bigger and stronger, they sometimes intimidate fillies before the race even begins. But many experts were picking Ruffian in this rare match race: she was taller by three inches, heavier by sixty-four pounds. Even Vasquez, who had ridden Foolish Pleasure in his last nine races, switched back to Ruffian for the match race. In a prerace workout, the *Daily Racing Form*'s chief clocker said of Vasquez and Ruffian: "I think if he turned her loose my watch would explode. I've been around about half a century and I've never seen a Thoroughbred work so fast so easily." Foolish Pleasure was just as impressive. Anticipation for the race was monumental: some twenty million people would watch on television.

Barbara Janney remembers a moment when Ruffian halted as she was led past the grandstand. The crowd was roaring — and this was *before* the race. Track aficionados would later say they had never heard a crowd so animated. Ruffian paused to consider these onlookers, as if certain she held center stage. Janney would savor that memory because of what followed. Maybe that is how she wanted to remember Ruffian: composed and self-assured, acutely aware of her own greatness.

When the gates broke open, Foolish Pleasure smartly took the lead on the outside, but within several strides Ruffian had nosed out in front by a few inches on the inside. She was so much bigger than he was that fans in the stands could not see the colt as the two horses rounded the clubhouse turn for the backstretch. The filly looked to be running alone.

Near the halfway point, with Ruffian ahead by half a

length, it happened. Both jockeys heard a snap. It is the sound a hefty branch makes when it cracks in a windstorm. It is the sound of bone breaking, and those who have heard it even once never forget it or care to hear it again.

A Thoroughbred is a remarkable animal. Within six strides of exiting the starting gate, the horse is streaking at forty miles an hour and taking in five gallons of air a second. The force on the horse's front cannon bone has a calculated impact of ten thousand to twelve thousand pounds. The red line is perilously close: at eighteen thousand pounds the stress on the bone is simply too much to bear.

Ruffian had taken what the Thoroughbred world calls "a bad step." That curious expression appears to blame the horse or the Fates for what is often a numbing, life-ending event. Her right front hoof, suddenly loose and flapping, no longer supported the leg. She was running on raw bone. It splintered, one vet said later, like an ice cube hit with a hammer.

At the moment the leg broke, Ruffian appeared to bump Foolish Pleasure. But the videotapes reveal that she was actually *leaning* on him, trying to compensate, to remain on three legs and keep running. On she ran, somehow, for fifty more yards, while Vasquez desperately struggled to rein her in. Then she veered right and staggered to a halt, blood and bone issuing from the terrible wound. The jockey leaped off, held his hands aloft and tried to support Ruffian, who screamed in pain.

In the stands, disbelief turned to utter, open-mouthed silence. The roaring ceased the way a tap cuts a jet of water. Even the race announcer paused, as if he, too, could not comprehend what he was seeing. Some in the crowd began to weep — women and girls wearing T-shirts emblazoned with the name Ruffian and the circle-and-cross symbol of womanhood, men and boys who had a moment before cheered Foolish Pleasure's apparent surge.

By gamely continuing to run, Ruffian had made the injury many times worse. On the track, men linked hands under the stricken horse and lifted her into a horse ambulance.

Ruffian thrashed and threw off the temporary cast they later put on her, along with another following surgery at Dr. William O. Reed's equine hospital. Lying on her side on a padded floor, she came out of the anesthesia and tried to run — from the pain, perhaps, or to catch Foolish Pleasure. She spun round and round in a circle, like a butterfly with its body pinned to the ground. "The same thing that made her win," said her vet, "made her die." Burly hospital staff trying to restrain her were tossed about as if they were made of balsam.

Finally, the question was put to the distraught Barbara Janney: do we operate again? And the answer came back: end her suffering. In a rare and private ceremony at Belmont the next evening, they buried Ruffian. She was bound in white like a mummy, and a great hydraulic lift gently laid her in her grave, twelve feet deep and square, her head pointing toward the finish line. Looking on were the grief-stricken Whiteley, exercise riders, stable hands, Vasquez in a dark suit. An assistant trainer put on Ruffian two blankets she had worn, along with some flowers. The trainer fussed with the blankets, made them smooth, then the machines covered her with earth and a horseshoe wreath was laid atop her grave.

At the park there stands a stone marker listing Ruffian's victories. And sometimes people send flowers to Belmont with instructions that they be placed by the obelisk. The senders no doubt remember, with sadness and with joy, how the beautiful black filly erected sails between those long, elegant strides and rode the wind at her back.

Great horses are "freaks," say trainers. They use the term with affection to describe horses whose physical gifts, courage and intelligence combine to let those horses do what they do so well. In the case of Secretariat and Phar Lap, the gift was a literal great heart, an uncommonly big and powerful engine buttressed by other intangibles the horse world ends up calling class or spirit.

I *like* the fact that — so far, anyway — you cannot predict or orchestrate such physical genius.

Conformation faults aside, it is still true that every foal out of no-name sires and dams has promise. Secretariat's high-priced progeny, meanwhile, carry no guarantees. Certain Thoroughbred lines, such as those of Eclipse and Northern Dancer, generate more of their kind. But not always. In any one mating, the mare matters as much as the sire. In any given race, longshots can come in. Jim Elder won a gold medal in show jumping for Canada at the 1968 Mexico Olympics on a six-year-old horse, Immigrant, plucked from a New Jersey riding school. Of the three other horses on that team, one was a Thoroughbred bought off the track for $500; another had been spotted in a field.

As ruinous as the track has been for countless trainers, riders and owners — most of whom must win, after all, to live well — for the rest of us without a serious stake it can be a place of hope and surprise. I do not bemoan the five bucks gone when my longshot drifts home last, but when the horse comes in I am over the moon. Some horses *always* come in.

No listing of great horses, however brief, would be complete without the magnificent English chestnut who never lost a race. Almost seventeen hands high, Eclipse was unusual in his conformation, with hindquarters an inch higher than his withers and a long slim neck. Paintings of the horse look bizarre, as if an amateur had gotten the proportions all wrong.

Winner of at least eighteen races before he died in 1789, Eclipse sired the greatest line of winners the world has ever seen: of the horses who won 170 major races around the world in 1979, 82 percent could be traced back to Eclipse. Northern Dancer may one day eclipse Eclipse, but until then the colt born during an eclipse of the sun will remain king of Thoroughbred sires.

Eclipse was first owned by a duke, then bought by a sheep farmer unable to handle him — the farmer, in fact, gave serious thought to gelding him. At the age of five Eclipse was leased to an Irish-born army captain named Dennis O'Kelly and only then did his racing career begin. His first race was at Epsom, where he easily won the first of his two four-mile heats.

Emboldened by beer at a local inn, O'Kelly boasted he could predict the order of finish for the next day's four-horse heat. When someone called his bluff, he wrote a note and asked that it be read aloud after the race. Eclipse won next day by a quarter of a mile and O'Kelly's words earned a place in the lore of racing. "Eclipse first," he had written, "and the rest nowhere."

I chanced upon a book called *The Horse, With a Treatise on Draught; and a Copious Index*, Published Under the Superintendence of the Society for the Diffusion of Useful Knowledge in London, 1821. The musty old book referred to Eclipse as "a thick-winded horse" who "puffed and roared so as to be heard at a considerable distance." Some people had heard about the horse but arrived moments too late for the Epsom trial. They did, though, encounter an old woman who reported seeing "a horse with white legs running away at a monstrous rate, and another horse a great way behind, trying to run after him, but she was sure he would never catch the white-legged horse if he ran to the world's end."

Horses are the most valuable animals on the planet — her owners tried to save Ruffian out of sympathy and affection, because she was Ruffian and because her worth as a breeding mare was almost incalculable. Consider, too, the finest race-horse ever bred in Canada: Northern Dancer. In the early 1980s, his stud fee was $300,000 and rising. At forty mares a season, his annual fees neared $12 million. At the peak of the buying frenzy, one of Northern Dancer's yearlings (Snaffie Dancer — a dud, it turned out, as racer and stud) sold for a world record $10.2 million. In the spring of 1982, Northern Dancer's owners spurned an offer of $40 million from a French syndicate. "There are some people," Charles Taylor, the horse's owner, said then, "who believe that everything has its price — but not Northern Dancer." In his lifetime, the stallion sired one thousand foals.

He was a horse of immense character and lordliness. If breakfast came late, he would toss his head impatiently and kick the side of his stall. Charles Taylor's mother used to bring the horse sugar cubes, and when she forgot he nipped her. He bit his handlers and tried their patience and was almost — almost — gelded.

Bernard McCormack, the general manager at Windfields Farm north of Toronto, where the gifted colt was born, talked to me in his spacious office where the light pours in and mementos of Northern Dancer abound. The horse, he said, had a strong, individual temperament right from the start. "He knew he was important. He stood out as a foal. Sometimes foals in the paddock won't follow the mare. Even at five days they're dictating where the couple will go. The foal leads! When you see foals that young with that kind of presence, by and large they have ability."

Later, at stud, said McCormack, Northern Dancer had, if anything, an even greater sense of his own worth: "He had an ego the size of Mount Everest. When people came to see him he would pose for the cameras."

Northern Dancer came in from the paddock only when he felt like it. "He would make you wait," said McCormack, "and you wouldn't go out to the field to bring him in because he'd run you over. But he was Northern Dancer and you make allowances for a horse like that."

Peter Gzowski, the Canadian broadcaster and author, wrote a book in 1983 on racing, wherein he coined a neat phrase about Northern Dancer as stud: "Somehow, he was able not only to pass on his racing ability to his sons but also to pass on, as it were, his ability to pass it on." Breeders call the ability of a stallion to pass on his quality *stamping the get*. The Dancer stamped the get's get.

"It's a story without an end because it's still being written," said McCormack. "For that bloodline to dominate into

7.3 Northern Dancer holds off Hill Rise in the dramatic finish to the 1964 Kentucky Derby.

a third generation of horses is *the* most remarkable thing about him. The rate of fade is the slowest of any great stallion we've ever seen. Three and four generations of dilution and it's still going strong. His blood will remain a cornerstone of the breed and in time 50 to 70 percent of all Thoroughbreds will carry his blood."

They say of Northern Dancer what they say of only the best racehorses — "He ran a hole in the wind." He was born at fifteen minutes past midnight on May 27, 1961. As a fifteen-two-hand adult horse he was somewhat short for a Thoroughbred, but his stockiness and bull neck meant no one could call him small. In color he was a bay, with three white stockings and a thick diagonal blaze that ran from his fore-lock down his long head and into his left nostril.

His sire was Nearctic; his dam, Natalma (daughter of the great Native Dancer), and E. P. Taylor, Charles's father, made the colt available to buyers at eighteen months for the sum of $25,000. Every year, Taylor offered half the crop of Thorough-breds born at Windfields. One prospective buyer took the colt to his barn for a closer look . . . and brought him back.

The colt quickly earned a reputation as "feisty, willful and Napoleonic." Among the exercise riders he was known as "a very bad ride." Joe Thomas, then manager of Thoroughbred operations at Windfields, remembered him this way: "The low lad on the totem pole had to ride him and if he didn't use all his expertise, [Northern Dancer] would put him on his butt. Even at the track he would 'do tricks,' bolting, buck-jumping, all sorts of things. He was a handful." Like many fine athletes, he never got up for practice, just the games.

Northern Dancer was a horse with attitude, but how he could run. Like a fullback with breakaway power. He had his own style, not the classic Thoroughbred flow and grace, but a choppy, churning gallop. "A vest-pocket Hercules," the *Daily Racing Form* called him.

When Northern Dancer stepped onto the historic turf of Churchill Downs in Kentucky on May 6, 1964, I was a boy of fifteen. I remember the patriotic stir this horse provoked. It seemed like all of Canada watched the race that day. In 1919, Sir Barton had become the first Canadian-owned (not bred) horse to win the Triple Crown of racing, but no Canadian-bred horse had ever been given much of a chance in the Derby. Northern Dancer, then, stood to make history.

Willie Shoemaker, the legendary jockey who had been riding the colt, switched for the Derby to a California-bred horse named Hill Rise, winner of his last eight races. Shoe-maker said the Dancer felt "rubbery" after longer races and would not go the Derby's mile-and-a-quarter distance.

Horatio Luro's instructions to Bill Hartack — a jockey known for being aggressive — reflected what the trainer knew of the horse's spirit. Keep him off the pace for about three-quarters of the race. Move on the far turn, if possible. Give him encouragement, never punishment. When he is hit, warned Luro, "it can turn sour."

Hill Rise went off as the six-to-five favorite; Northern Dancer was the second choice at seven-to-two. Hartack wore

Windfields colors: a blue silk jacket with gold dots on the sleeves, and gold silk over his crash helmet. The Dancer, four inches shorter than Hill Rise, wore number seven.

Out of the gate Northern Dancer settled into sixth, but by the clubhouse turn he was boxed on the rail. Boldly, Hartack took him to the outside and used the whip. But it was little more than a tap and the horse moved into the lead by two lengths, with Hill Rise in pursuit. I have watched that race over and over again: Hartack hit Northern Dancer ten times, always on the left to keep him off the rail. Hill Rise edged forward, and it seemed he might catch the feisty Dancer, but the Dancer hung on, winning, as one announcer put it, "by a long head."

Northern Dancer would go on to win the Preakness but not, alas, the Triple Crown. At Belmont he came up short, or maybe the jockey held him back too long in that grueling mile-and-a-half race.

Northern Dancer's time in the Kentucky Derby was a scorching two minutes flat, a new record for that race. It was faster than the eighty-nine Derby winners before him and faster than the thirty-four since — with the exception of the immortal Secretariat, who eclipsed that time by three-fifths of a second. Northern Dancer would win fourteen of his eighteen races and was never out of the money.

One day in November 1996, as the rain turned to sleet, I stood alone by Northern Dancer's grave at Windfields. A sadness hung about the place. The farm that had produced more stakes winners than any in the world was closing down its breeding program; clients could still breed horses at the farm but only their own horses. Colts would no longer bear the Windfields imprimatur of quality. Charles Taylor was ill (he would die of cancer the following year) and a sixty-year-old era was winding down. (In 1998, Windfields did rekindle its breeding program but in a very modest way.)

I stared at the cold gray granite marker that bore Northern Dancer's name and dates, sire and dam. Nearctic-Natalma,

1961–1990. A rectangular perimeter of straw protected a bed of roses that would bloom in summer, the same red roses they put around his neck after the Derby. Reluctant to leave, I shivered in the wind and greeted War Deputy, a stallion in a paddock close by. He seemed keener on prancing and baring his teeth than on any show of cordiality. No surprise. He had Northern Dancer blood in him. The best.

Behind me was the foaling barn — crowned by a black weather vane of stallion, mare and foal — where the Dancer was born. Stall number two lies about thirty yards from his grave.

When the great Secretariat died on October 4, 1989 at 11:45 in the morning, his body was conveyed immediately to the University of Kentucky, where Dr. Thomas Swerczek, a professor of veterinary science, performed the autopsy.

"We were all shocked," he said afterward. "I've seen and done thousands of autopsies on horses, and nothing I'd ever seen compared to it. The heart of the average horse weighs about nine pounds. This was almost twice the average size, and a third larger than any equine heart I'd ever seen. And it wasn't pathologically enlarged. All the chambers and the valves were normal. It was just larger. I think it told us why he was able to do what he did."

Secretariat was special. Ron Turcotte, the Canadian jockey who rode the chestnut stallion many times, remembers being approached before the Belmont in 1973 by the venerable Hollie Hughes, trainer of a Derby winner fifty-seven years beforehand. Hughes looked at the horse, glanced at his workout times and calmly told Turcotte, "Son, you're riding the greatest horse that ever looked through a bridle. I have seen them all, including Man o' War. Secretariat is the best." Hughes offered Turcotte a prediction (You'll win) and advice (Don't fall off.)

Some called Secretariat the perfect racehorse. Even the so-called objective voice of science ran out of superlatives in

7.4   Secretariat racing: "He looked like he would run through a stone wall."

trying to describe him. George Pratt, an MIT scientist and authority on the biomechanics of the equine gait, examined Secretariat at Claiborne Farm and could scarcely believe what he saw: "He looked like he would run through a stone wall. He is a mountain of muscle, a mountain of dignity, a mountain of aristocratic bearing — the most impressive live creature I have ever looked upon."

Only the comment about aristocratic bearing missed the mark. What endeared Secretariat to so many was his playfulness. Turcotte often visited him at the farm after the horse had been retired to stud, and he was certain Secretariat recognized him. The still imposing horse would gambol across the paddock, stop at the fence and stick out his tongue. Turcotte would shake it and say, "Hello," and the horse would then rub his face against his old jockey.

William Nack, a writer who followed Secretariat's career with particular passion from beginning to end, spent forty days hanging around Secretariat's stall, talking to grooms, trainers and likely the horse himself. One day Secretariat grabbed Nack's notebook with his teeth but dropped it agreeably when the groom asked him to. Another time, after the groom had raked the shed, Secretariat got everyone in the barn laughing when he grabbed the same rake with his mouth and commenced to pull and push it across the floor.

"Secretariat," Nack wrote, "was an amiable, gentlemanly colt, with a poised and playful nature that at times made him seem as much a pet as the stable dog was." Yet he also loved to work and needed hard, fast workouts before a race to burn off the fifteen quarts of oats he ate every day during the season he won the Triple Crown.

Turcotte fondly remembers the horse's great intelligence, how inclined he was to ham for photographers, how manageable and generous in a race, how handsome, how powerful and yet how kind — "as sweet as a lamb." He was a horse who loved humans.

But as is often the case at the track, only numbers give the true measure of Secretariat. Here are some to ponder: 25 $^1/_5$; 24; 23 $^4/_5$; 23 $^2/_5$; 23. These were Secretariat's quarter-mile splits when he won the Kentucky Derby in 1973, coming from dead last and accelerating in each successive quarter mile en route to the fastest Derby time ever. He then won the Preakness, again in record time (though a malfunctioning timer failed to make it official), before winning the Belmont by thirty-one lengths.

Finally, there is this number: ten thousand. That many pilgrims a year used to go to Claiborne Farm to see Secretariat and pay homage. A farm owner observed that Secretariat was no longer a horse; he had become a legend. But by October 1989, the eighteen-year-old horse had laminitis, a life-threatening hoof disease and his prospects were bleak.

The stricken Nack had seen him grazing in the distance the day before and wanted to look upon him one more time. But Secretariat had taken a bad turn, and the request was denied. "Remember him how you saw him" advised the horse's owner. The next morning at dawn, Secretariat lifted his head and nickered loudly — "Like he was beggin' me for help," his groom would later say. Before morning was out, the great heart would cease to beat.

There have been other big horses, other big hearts. Think of Phar Lap, the New Zealand-bred horse also known as The Red Terror, Red Lightning and Big Red. The name means, in the language of Thailand — or Siam, as it was called in his day — lightning or "wink of the sky." After he died, his heart was preserved and displayed in a museum in Canberra, Australia. For comparison's sake, curators parked it beside the heart of an Australian cavalry horse and listed the two weights. The remount's weighed seven pounds; Phar Lap's weighed more than thirteen.

In the Great Depression of the late 1920s and early 1930s, Phar Lap, "the wonder horse," provided almost as much diversion as the dimpled Shirley Temple. He was a seventeen-one-hand, 1,250-pound chestnut horse who allowed a toddler to ride on him but would buck off experienced jockeys when they dared order him about.

A brilliant horse inclined to mischief, he would grab grooms by the shirt, tug sharply and whinny gleefully when the shirt ripped. Phar Lap adored his personal groom, Tommy Woodcock, who always had a cube of sugar for him and called him "Bobby Boy." After a painfully slow start to his career, he won thirty-seven races in three years. "He's not our horse," his trainer's wife once remarked. "I think every child in Australia owns him."

The public idolized him, but not so the bookies who lost their shirts faster than Phar Lap's first grooms. Anonymous phone calls issued threats: acid in the horse's face, poison in his feed. A passenger in a car once leveled a double-barreled shotgun at Phar Lap, but Woodcock put himself between the gun and the horse and somehow the pellets missed. Race officials kept adding more weight to handicap Phar Lap, and though he continued to win, his owner — fed up with all the harassment — finally took him to America.

His jockey, Jim Pike, likened trying to hold him back to

stopping a freight train. A Sydney vet, Dr. Stewart McKay, observed of Phar Lap that "his muscles are such that his hind legs should be cast in bronze so future generations can bow in admiration before such perfect propulsive machines."

Phar Lap died in California suddenly and mysteriously, his head in Woodcock's arms, his body swollen like a balloon. Perhaps significantly, the threats that had hounded Phar Lap out of Australia had resurfaced in America. If the cause of death was colic, why did such a strong horse succumb so quickly? If sour feed killed him, why was no other horse in the stable affected? Was Phar Lap the victim of an antigambling zealot?

The postmortem reported that the horse's stomach and intestines were severely inflamed — "as if from an irritant poison." Subsequent examinations contradicted that finding, and the hasty removal of his heart to the museum may have stymied further, perhaps conclusive, analysis. The words of a jockey who rode against him may well have been prescient: Phar Lap, he complained, was *too* good. "Sure things" upset the system — a sometimes nasty underworld built on chance and greed.

Last but not least of the speed-horse legends is another horse they called Big Red around the barn. Everything about Man o' War was big. His appetite. His size. His reputation.

Born March 29, 1917, at the height of the First World War (and thus his name), Man o' War quickly became a national hero in the United States and the focus of spectacular bids to buy him. One Hollywood movie producer dangled $1 million in front of Man o' War's owner. The Texas cattle baron and oilman W. T. Waggoner signed a blank check and invited the horse's owner to name his price. But Man o' War was not for sale.

He seemed to inspire affection in everyone who saw him run. His workout times were often faster than his race times, which were themselves stunning. Man o' War ran for the pure joy of it, quickly staking out his position at the front and defying any horse to catch him or even to get close.

He lost only one race: a poor start left him ten lengths behind, but he easily recovered and could have won had he not been boxed on the rail. The winning horse that day was called, fittingly, Upset, and his jockey, Willie Knapp, later wished he had just moved over and let the legend breeze past. "So great a champion as Man o' War," he said with feeling, "deserved to retire undefeated."

Man o' War won one of his twenty-one races by one hundred lengths and set record times (three world, two American, three track), not by a fraction of a second but by six seconds or more. Save for that one loss, he was never pressed.

The almost sixteen-two-hand chestnut had a huge stride and the manners of a blueblood. All his life he raised a fuss about being saddled, yet he was playful, too: like an obedient dog, he would carry his groom's hat in his mouth.

Next to running, Man o' War loved eating. His handlers sometimes kept a bit in his mouth, hoping to check his habit of hastily devouring huge quantities of food. Every day Man o' War ate all the hay he could eat, twelve quarts of oats and a handful of carrots. As a racer he weighed a hefty 1,150 pounds; as a breeding stallion he shot up to 1,370 pounds.

Even as a stately twenty-five-year-old horse, Man o' War never stopped acting the champion. He had to be locked in his stall to prevent him from challenging young stallions in adjacent pastures to impromptu races. And so Big Red is remembered: the mane flying, tail high and proud, the stride long and easy.

In October 1989 I was in Stuttgart, Germany, watching Ian Millar ride the incomparable Big Ben in a grand prix event. For Ian's memoirs, I wanted to walk a little in the lanky rider's shoes on this European tour. After many hours at his farm near Perth, Ontario, I was acquiring a sense of the man. I was also feeling more and more affection for his striking chestnut horse, the Thoroughbred-Belgian warmblood cross then dominating the world of show jumping. Big Ben had won double gold at

7.5 Big Ben led by groom Sandi Patterson: lord of the show ring every
    time out.

the Pan-American Games two years previously and had done
what no horse had — taken the World Cup two years in a row.
"Brave — brave — brave," Ian once said of him.

Big Ben entered the show ring with princely bearing, tail
and head high, supremely confident. A frisson of excitement

rippled through the audience even before his name was announced.

But Big Ben was a true freak: a seventeen-three-hand horse by a small stallion out of a tiny mare, neither of whom had ever produced offspring even approaching in size or merit that one glorious colt.

Big Ben had his idiosyncracies. He loathed plastic bags and the sound of trains, loved his routine 10 A.M. ride. If it came late, his anger showed. When his devoted groom Sandi Patterson took a rare day off, he was surly with her stand-in and pouted for days. But Big Ben was coltish, too, and would steal Sandi's shoes and toss them in the air when she lay on the grass beside him.

I remember my time at Millar Brooke Farm fondly, and especially one summer day watching Ian and Ben glide over the outdoor course, me leaning against a tree in the company of sundry dogs and cats. The loose beauty of their partnership left me feeling a small sense of privilege, for I was an audience of one. I might have been in a bar at noon as Miles Davis and John Coltrane rehearsed or in a cavernous arena as Bobby Orr and Gordie Howe played shinny.

When we first arrived in Stuttgart, photographers and journalists swarmed Big Ben's stall. The lord of the ring was a favorite every time out.

Grand Prix day is typically on a Sunday, and this horse had a sense for Sundays. He bobbed his head and pawed the ground as Sandi tacked him up that morning in Stuttgart. Competing against forty other horse-and-rider combinations, Ian and Ben went clear in the first two rounds. In the jump-off they faced three world-class riders along with Nick Skelton of Great Britain — maybe the finest speed rider in the sport.

Ian and Ben's time was a blistering 32.76 seconds, well under the forty-seven seconds allowed. No other rider matched that time or even went clean.

As is the custom, during the awards presentation all the lights in Hanns-Martin-Schleyer-Halle were turned off and everyone in the stadium flicked on the cigarette lighters handed out earlier. Outside the stadium minutes beforehand, Ben had been pawing the air, Trigger-like, when police-band tubas warmed up next to him, but at that moment he was standing calmly in the center of a pitch-black ring, the spotlight on him and the yellow first-place ribbon at his shoulder.

Capturing that magical moment is a photograph: background velvety black, so black that Ian's riding helmet seems to disappear into it, with only a small patch of brown turf showing in the bottom left-hand corner. The photograph has the feel of a portrait painting. Horse and rider exude haughtiness: Ian eyes the camera, a tight, proud smile on his lips; Ben, ears pricked forward, looks slightly to the right but the intensity is discernible in his visible eye. There is no hint of the filled stadium, only the glow of this one victory, and the promise of more.

The glory of Stuttgart, and indeed that entire European tour — which saw Ben receive a hero's welcome in Belgium, where he was born — would be followed by three calamitous events: two colic surgeries a year apart and a horrific highway accident in 1992 that killed one man and one horse and badly injured several others. The wonder is that Big Ben rebounded.

Two weeks after the accident, to the astonishment of the show-jumping world, he won the grueling derby at Spruce Meadows in Calgary — for an unprecedented fifth time. One of his last great victories, in 1993, was that same derby. The record will never be matched.

In 1994, during the writing of *Big Ben*, a book for younger readers, I talked at length to Sandi Patterson, who had to relive the surgeries and the gruesome accident on the prairie. For a time, one image haunted me: the broken horse van on its side, the barefooted Sandi on top, screaming Ben's name into the dark vault as the rain pelted down. His only groom for seven years, she had come to love him. Sandi taught me a lot about the powerful bond that can form between horse and human. I confess I found it odd and mystifying then. Not anymore.

On Big Ben's farewell tour in the summer of 1994, people lined up for hours under a hot sun to touch him or be photographed with him. They cherished the horse, not just because he won so often in his spectacular ten-year career, but because he won against adversity. His size, his contrariness, the colic, the accident — all could have stopped him. He had heart, and anyone who ever saw him jump a fence sensed it.

Compared with racing, show jumping is a young equestrian sport. The term first gets a mention in French cavalry manuals of the late 1700s, but only in the 1860s did organized show-jumping competitions get under way. Though early show jumpers rode ever so slowly and carefully — riders were then allowed all the time they needed — the sport took off. By 1900, jumping competitions formed part of the Olympic Games and international show jumping had been launched.

A classic story from the world of show jumping involves a gifted German rider named Hans Winkler and his horse, a notoriously difficult bay mare named Halla. She, like Ben, was a freak horse.

Sired by a German trotter and one-quarter Thoroughbred, Halla was first tried as a steeplechaser and then as a three-day event horse of possible Olympic caliber. But she proved too fiery for the dressage phase. To land, finally, in Winkler's hands seemed a stroke of luck, one that makes some of us believe in fate or destiny. If Ian Millar and Big Ben were the perfect equestrian match of the 1980s, Halla and Winkler were their counterpart in the 1950s.

As rider and trainer Winkler had his own style, based not on the strict discipline of the German tradition, but on a keen insight born of psychology and diplomacy. Vicki Hearne, an astute writer and horse trainer, has called Winkler "one of the most tactful and reassuring riders who ever lived. I never saw anyone so light on the reins and in the saddle."

Winkler knew he had a special horse. Though tall, she was nimble, and despite her leanness, she was an extraordinary jumper who won numerous puissance competitions. She was also uncommonly courageous. Proof came in the 1954 world championships (which required that competitors ride one another's horses). Halla, a nervous horse made more so by the many new riders, had one disastrous twenty-six-fault round. "Crashing and burning," riders call it when numerous fences go down, often dramatically.

Though she came up lame, and though the horrified Winkler begged the judges to let her withdraw, the rules of the day obliged her to cover the course a second time with the same rider. Other horses might have refused. Halla did what was asked of her and had only one fence down. Winkler — who won that world championship, by the way — knew then what a mare he had.

At the 1956 Olympic Games in Stockholm, Winkler pulled a groin muscle while vaulting the penultimate fence in the first round. Despite the pain, he asked Halla to take the last fence, which she did. Still in terrible pain, Winkler had to be lifted up into the saddle for the jump-off. Some horse-wise people thought it was madness for a rider to face huge fences on a high-strung mare when he could do little more than point the horse at the jumps.

Vicki Hearne later applauded both Winkler's trust in his horse and the horse's achievement. "Imagine a race-car driver," she wrote, "suddenly unable to operate clutch, brake and accelerator except in a distant, awkward and weak fashion."

In show jumping the takeoff point is critical: it is a matter of inches. If the horse leaps too soon or too late, if the horse momentarily loses focus — diverted by a banner or a camera's flash — the rail falls. Overly fast or overly slow horses will not manage a jump-off course. Turns can be neither too tight nor too round.

But in this game of high-speed chess it is the rider, not the horse, who calls the shots. Suffused with adrenaline, the rider must also allot parts of the brain to analysing the sequence of jumps, and number of strides between them, the seconds allowed and riding itself. When Winkler rode Halla that day, however, he vaguely hoped the horse could call the shots. Even to embrace such a hope, Winkler had to have trained her well. As Holly Menino put it in her thoughtful book on horse-human athleticism, *Forward Motion*, "Teaching a horse to jump fences is more like instructing a child to read than like training a dog to stay or to fetch." Only a literate horse could do what Halla did.

"All I had to do," said Winkler, "was just sit there doing nothing — she worked everything out." (Nor has she been alone in that ability. The steeplechase world was stunned six years after Halla's feat when a champion horse called Mandarin snapped his bit early in a race and his jockey rode him the remaining three miles without a bridle. He won — by a neck.) Other riders have pointed out how delicately Winkler sat the horse, and how terrible the pain must have been. "Doing nothing," certainly in this instance, was a feat in itself.

Halla won an individual gold medal for Winkler and team gold for Germany. When Winkler was helped out of the saddle, he did the right thing: he threw his arms around Halla and thanked her.

The Irish have a special feel for the horse, perhaps literally so.

Among the many legends that surround the aforementioned Eclipse and Dennis O'Kelly is one in which the latter is fleeing English law (for leveling English fences) and hiding in a stable. His quest is to breed the perfect "harse" by first finding the perfect sire and the perfect dam. In the pitch-black stall, he runs his hands over the horse near him and, upon feeling her lines and conformation, realizes he has found her.

A novel called *O'Kelly's Eclipse*, written by Arthur Weiss in 1968, speaks to the notion that fine horsemanship is an Irish birthright. "It was the limestone subsoil," claimed O'Kelly, "which made Ireland the best place in the world for the raising of God's chosen animal." There is truth in that: minerals in the limestone come up through the grass and are said to enhance bone development. Add fine rains to almost perfect pastures and, sometimes, you get almost perfect harses. Arkle was one.

He may well have been the greatest steeplechaser of all time. He was born on April 19, 1957, in County Dublin, where a disabled widow named Mary Baker and her son Harry kept up the family tradition of breeding a few steeplechasers.

Arkle, a gangly bay with large eyes and almost mulish ears, was a natural jumper who could, went the boast, leap the mountain in Scotland that was his namesake. The duchess of Westminster, an aristocrat of Irish origin, bought him at the urging of a trainer who lived near the Bakers and who steadfastly refused to rush young horses into competition. Like Eclipse, Big Ben and many other successful horses, Arkle enjoyed his youth and came to work relatively late in life.

Brave and determined as a racer, Arkle was also an unusually sociable horse, kind to children and friendly to dogs. His favorite meal, served at 4:30 P.M. every day, consisted of a bran mash mixed with six eggs, oats and — that crowning touch — two bottles of black beer. He was an *Irish* horse, after all.

For all his playfulness in the paddock, though, on the race course he was unmatched. His twenty-five victories included three consecutive Cheltenham Gold Cups. He always saved enough for his patented run at the finish. Sometimes Arkle ignored the finish line altogether and simply kept on running. Because he beat other horses so badly (by up to thirty lengths) and because a few horses died trying to catch him, some called Arkle a "killer" horse.

In his last race, in 1966, Arkle hit a guardrail and cracked a bone in his foot. The miracle was that he finished at all, let alone a close second. It must have caused him great pain to run, and many wept to see him limp back to the unsaddling ring. When he finally died in 1970, put down after arthritic lesions developed in both feet, the Irish mourned his passing as they would a head of state. "We will never," said his jockey, Pat Taaffe, "see his likes again."

Dangle gold bullion in front of a thief; walk a horse of high pedigree past him. The effect is the same. Madness sets in. The glories of the sport-horse world are many, but there also exists a dark and sinister aspect that may be inevitable when horse-flesh commands the price it does. This, too, is part of the mix, part of what the horse stirs in our hearts.

You may not know these horses, but I feel compelled to name them. Instant Little Man. Belgium Waffle. Rub the Lamp. Rainman. Roseau Platiere. Charisma. Streetwise. Empire. Condino. All killed on the orders of unscrupulous owners, trainers or agents, sometimes as part of a high-stakes insurance money grab and sometimes by a hired assassin named Tommy Burns, who came to be known as The Sandman.

In the late 1980s, Chicago attorney Steven Miller, who was trying to fathom the disappearance a decade earlier of a candy heiress named Helen Brach, opened a true can of worms. His investigation led to a horseman named Richard Bailey, who had courted the rich widow, and from there the net spread to include top-ranked show jumpers and trainers. Of the almost two dozen people charged, all but one was convicted and many served lengthy jail terms. The show horse industry's "dirty little secret," as a Miller colleague put it, involved "a virtual who's who of the nation's equestrian world."

That world clearly was rocked by this case, not the first of its kind. Twenty-five years ago, horses at prominent Thoroughbred racetracks in the U.S. were dying so frequently that insurance companies refused to insure the animals. Cynical grooms at Belmont Park would wonder aloud in the morning, "Anyone die last night?"

The history of racing is pocked with stories of murdered horses. For Bay Jack, the winner in 1869 of the Queen's Plate, the end came during a competition in Strathroy, Ontario. A competing jockey, intending perhaps to check Bay Jack's speed for a forthcoming race in Toronto, had given him too much laudanum. The death of the bay colt, described as "a noble, beautiful looking animal . . . a model of equestrian grace and beauty," occasioned a day of mourning in Strathroy. As one witness put it, "Had the man charged with administering the drug been caught he might have been lynched, so great was the feeling in the matter."

There is a joke in the sport horse world that goes like this: to compete at the highest levels you need two things — money and more money. After 1986 in the U.S., you needed yet more money. Tax changes prohibited horse owners from writing off losses. A million-dollar horse aging, underperforming or gone lame no longer constituted a depreciating asset but rather an expensive liability. For some owners not burdened by a conscience, the solution lay in calling The Sandman.

Tommy Burns had tried stuffing Ping-Pong balls up a horse's nose, then suffocation using green garbage bags, before he settled on electrocution. The method was simple and quick. He would cut an extension cord down the middle, then attach an alligator clip to the exposed wire at each end. He would then fasten one end to the horse's ear and the other to the horse's rectum before plugging in the cord. The dead horse would fall like a collapsing tower. His intestines would rupture, as they do when a horse dies of colic, and it took a careful and suspicious investigator to notice tiny burn marks on the horse's ear.

Burns confessed to killing twenty horses this way over the course of ten years, earning anywhere from $5,000 to $40,000 per execution. In his wake, some insurance companies have withdrawn from the horse business entirely; others remain on full alert. And well they should when people pay millions for a horse.

The Canadian show jumper Jim Elder worries that these vast amounts of money tarnish the sport. "These days you have a rider, trainer, buyer — all these middlemen. When so much money is involved . . ." Elder, a lifelong rider-trainer-buyer who has never paid more than $5,000 for a horse, does not finish the sentence. He just shakes his head.

"I never rode a horse I didn't like," he says, sounding a

little like Will Rogers. At sixty-four he still feels boundless enthusiasm for horses. "Every horse has his niche. They're all good for something — hunting, hacking, jumping." A place in the universe for every horse seems a far cry from The Sandman, whose clients would sometimes ask him to kill a horse and then, months later, the replacement.

An elite horse is a vulnerable horse, prey to all kinds of menace short of murder. At the height of Big Ben's career, his groom used to sleep in his stall at major horse shows, so great was the fear that someone would spike his food, which would have led to a doping charge against Ian and the elimination of the horse from the event or worse.

In 1991, Monty Roberts was in Germany working with a brilliant speedster called Lomitas — the top-rated Thoroughbred racehorse in that country's history — who balked at starting gates. Monty resolved that problem and helped spirit the horse away to California when blackmail letters threatened harm, but not before the horse was poisoned with a heavy metal — just enough, one letter had warned, to make him sick. Once again, a winning horse — "a freak" — paid the price of being better than the rest. More than ever, it is dangerous for a horse to be splendid, or splendid and suddenly not so.

I think I understand why Ruffian and others of her stature can move us to tears. To ride, even to watch, an elite sport horse is to feel the bearing and power, the *nobility* more than anything, that flows from a quality horse.

Ron Turcotte once observed that Secretariat had become a heroic figure in the time of Watergate, that shattering blow to our trust in high office. The timing, he believed, was no coincidence. There is something to his insight. The bloated salaries and egos of professional athletes have bankrupted a once reliable source of heroes, so the horse — a four-legged athlete with no desire to take drugs, snub reporters or hold out for extra millions — nicely, and majestically, fills the bill.

CHAPTER 8

EPIC RIDES

I know that [Mancha and Gato] understand the man
who shared so many perils, hardships, hunger,
thirst and weariness with them, in the long pilgrimage
through the Americas that has placed them with the
immortals of the equine race.

R. B. CUNNINGHAME GRAHAM, IN HIS
PREFACE TO *Tschiffely's Ride: Ten Thousand Miles*
in the Saddle from Southern Cross to Pole Star

M Y ATTRACTION TO horses complements another I
have — to remote places and lost eras. I am not alone.
When tourism is the fastest growing industry in the American
West, when visitors to Montana jack up the price of land by
buying it (thus T-shirts reading "Montana sucks — now go
home and tell your friends"), when dudes outnumber cowboys in Alberta, then the cowboy way may indeed be imper-
iled by eastern zeal. But it all speaks, finally, to the pull of
horses, and it was The Horse that drew me to the Willow Lane
Ranch in foothills Alberta.

At that time, my four long days in the saddle during the summer of 1995, including two days on a cattle drive, constituted epic rides for me. My body said so. Only later did I begin to collect stories about the real thing.

If I hold in awe such riders as A. F. Tschiffely, Barbara Whittome, Butch Cassidy and Felix Aubrey — who have a place in what might be called the long-distance riders' hall of fame — it is not because I, like them, know what it is to ride a horse every day for weeks or months or years at a time across or down continents.

Aubrey almost rode himself to death in 1848 — to win a bet. Cassidy and the Sundance Kid, who fled Wyoming and a gaggle of U.S. marshals, rode to the very tip of South America before the end came in Uyuni, Bolivia, in 1909. Tschiffely rode the other way in the 1920s — from Buenos Aires to Washington, a two-and-a-half-year trek. Whittome crossed Russia on horseback in 1995 to prove a point.

And while the jocular *City Slicker* movies feasted on the clichés of the cattle drive, it is sobering to ponder the genuine article. A typical cattle drive in the nineteenth century once meant a thousand miles in the saddle — one way.

The first cattle trail in America spanned the distance from San Antonio, Texas, to Abilene, Kansas, and was carved out by Jesse Chisholm, a half-Cherokee cowboy. Other trails — the Sedalia, the Western, the Goodnight-Loving — would follow.

The longest cattle drive of all began in 1866 when Nelson Story sewed $10,000 into the lining of his clothes after finding gold in Montana. His idea — buy cattle in Texas, sell them at a profit in Kansas — seemed solid.

But two troublesome elements — Kansas farmers opposed to cattle tramping their land and cattle rustlers called "jayhawkers" — proved too much of a barrier. Then Story remembered how desperate for beef were the men of the Montana gold fields. He and his crew then headed cattle west along the Oregon Trail. But at Fort Laramie, the U.S. Cavalry warned him of hostile Sioux; at Fort Phil Kearny, which was virtually

8.1   At the end of long and perilous cattle drives, all paths led to the
       saloon.

under siege by Red Cloud's warriors, the fort commander
ordered Story to corral his herd. He could go no further.

Undeterred, Story bought Remington rapid-fire rifles for
all his men. Unless he pressed on, he would have to sell his
cattle to the army, but at nothing like the rates he could command on the gold fields. That night around the campfire,
Story took a straw poll among his men. All but one voted to
push on, and next morning, October 22, the lone cowpuncher
was forced at gunpoint to join the others. Story and his men
did indeed battle the Sioux and who knows how many other
obstacles along the way. On December 9, though, they arrived
in Virginia City, near the gold fields, with six hundred head of
cattle, four hundred less than they had set out with. They had
traveled almost two thousand miles.

We tend to romanticize cattle drives — the black coffee
round the campfire (cowhands used the butt of their guns to
grind up coffee beans), the stars and the sagebrush, the lure of
the open trail.

But the work was hard and conditions harsh: vaqueros, or cowboys (a term that originated in Ireland and only found common usage in Texas around 1900), ate a steady diet of dust by day and slept on the same dust at night. Semiwild longhorns were notoriously cantankerous and would use their massive horns (often five feet from sharp point to sharp point) to kill an unhorsed man. Swollen rivers, angry Indians, deserts and outlaws could any or all of them kill the cattle driver.

The diaries of Cyrus C. Loveland, who moved cattle from Missouri to California in 1850, are both memorable and telling. Provisions gone, living only on beef and having driven the cattle continuously for two days and two nights to cross the desert, he and his men look a ghostly crew. Here they are at the Truckee River, near present-day Reno, Nevada: "The last night on the desert we were so overcome with sleep that we were obliged to get off our horses and walk for fear of falling off. As we were walking along after the cattle it certainly would have been very amusing to anyone who could see us a-staggering along against each other, first on one side of the road, then the other, like a company of drunken men, but no human eye was there to see, for all alike were sleeping while walking."

Rare was the trail driver over the age of thirty. Most men quit after ten to fifteen years. Years of sleeping in the open on cold, hard ground, all the while ignoring sickness and injury, produced the "stove-up" cowboy — a man who faced ill health for the rest of his life.

The exquisite skills these men possessed were much admired, not least by cowboys themselves. "It was a pleasure to see them work," wrote John Clay in *My Life on the Range*. "They swept round a herd with an easy grace and careless abandon."

The cattle drive moved in somewhat hierarchical fashion. The trail boss took the lead, with his best men riding right behind at "point." Others farther behind rode "swing" and "flank." Bringing up the rear and eating the most dust of all were the "drag" riders, usually the youngest or least accomplished. Stampede was a constant threat, and at times the cowboys would force the cattle to circle, singing loudly to quieten them (thus the "singing" cowboy).

There existed a close fraternity among the men, who were governed by a code of conduct. Cowboys were expected to be cheerful even when sick or tired, never to complain, always to be courageous (sacrificing their lives for the herd, if necessary) and to help a friend — even an enemy — in distress.

Care of your remuda (Spanish for "replacement") of six or so horses per cowboy was part of that code. The hand who abused horses was sent packing. "An empty saddle," the saying went, "is better than a mean rider." When the trail boss assigned a new hand his horses and said nothing about their idiosyncracies, this was taken as a compliment to the new man, a vote of confidence in his ability to handle any horse. And if the boss wanted a man to quit, he simply took away that man's favorite horse.

Erwin Smith, at the age of eight, started work at the JSC Ranch near Quanah, Texas. He longed to become an authority on the Old West and in 1905 he set out with his box camera for some of the biggest ranches. His black and white photographs — remarkable for their quality and packed with detail — rank among the first and the finest to capture real life on those sometimes awesome ranches. (Richard King, for example, a steamboat captain turned rancher, founded a ranch in the late 1800s that boasted 1.3 million acres, a hundred thousand cattle and ten thousand horses.)

Ranch horses were often wild mustangs taken from the plains. They were short, wiry and tough, and even tacking them up was a battle, as several of Smith's photographs demonstrate. In a particularly telling one, a horse, though saddled, has decided not to be ridden: the mare is high on her haunches, her neck and head far back. A long rope leads from the bridle down to the cowboy on the ground. The rope is so firm yet relaxed in his grip he could be flying a kite.

8.2   Self-portrait of the legendary photographer of the old west, Erwin E.
Smith, 1908.

The photograph suggests that horses and riders were playing roles in a drama acted out daily. Indeed, Smith calls this
series of photographs *The Usual Morning Fight*. Any antics
that occurred around tacking might recur three or four times
a day — every time the cowboy changed horses.

Such spirited horses help explain the sizable spurs evident
on the cowboy trying to bring his horse down from the clouds.
Like most cowpunchers in Smith's photographs, this one wears
a Stetson, a high-crowned wide-brimmed hat that spoke more
of practicality than of fashion. The layer of air between the top
of the hat and the wearer's head offered protection from the
sun; the four-inch brim shaded his eyes. To top it all off, the
hat was waterproof.

But on cattle drives there was no real protection from the
weather. A nineteenth-century painting depicts a drover during
a fierce winter storm. He wears sheepskin chaps and he's bent
over miserably in the saddle, with both hands in his pockets, a

scarf over his mouth. The horse has icicles hanging from his nose and tack; his eyes are almost shut. Cattle even a few feet away are barely visible in the blinding snow. All look as if the wind at their backs has found a way into their bones.

One blizzard that hit South Dakota in 1886 blew for eleven days. Cowboys froze to their saddles. Like the men and animals in the painting, herds simply put their backs to the wind and drifted with it. Thousands of cattle died in this, the biggest "die-up" ever.

No blizzards or die-ups on my cattle drive. At the end of the day — a hard day, I thought — were clean sheets, a fine meal around the ranch table, even a hot tub for my rubbery thighs. For the first time in my riding life (until then, doled out hourly in a paddock or covered arena, maybe a trail), I would learn that saddle soreness is a minor term for major pain.

The Willow Lane, a working ranch ninety miles south of Calgary and run by Keith and LeAnne Lane, is small as Alberta ranches go: a string of twelve horses, almost three hundred head of Simmental-Hereford-cross cattle on 1,625 acres and a bed-and-breakfast-cum-vacation business to make ends meet. Twice a year, Keith, daughter Lyndsay, a wrangler and a few guests drive cattle from the ranch to a forest reserve where the wind-stunted limber pines grow. The twenty-four-mile cattle drive, in deference to ranch hands with soft hands, is stretched over two days. Real cowboys would do it in one, albeit a long one.

Keith Lane looked the part of cattle rancher, with a dark band of sweat around his gray felt cowboy hat, keen eyes (the yawning silver buckle on his belt proclaimed a trapshooting championship) and a walrus mustache I found myself staring at. Variously, the mustache and its tangled sideburn neighbors were, I concluded, the color of the setting sun, of ripened grain, of a red-brown heifer. The salt-and-pepper gray no doubt came from years of worrying that the price of beef would fall. Again.

When he doffed the hat, his face was revealed in all its two-toned glory. The top half, an egg-shaped hairless dome, was white, but below a demarcation line over the eyes, where the hat sat, there began a windblown, sun-beaten territory that ended abruptly at the collar. Decades of working outside will do that to a fella's forty-three-year-old face.

LeAnne Lane, formerly of Rosemary, Alberta, seldom rode, what with running a B and B. But I *knew* she could ride. In the living room her father, Reg Kesler, looks Marlboro Man-handsome on the GWG jeans poster that proclaimed him All-round Canadian Cowboy Champion of 1951. Lyndsay, the Lanes' fifteen-year-old daughter, rode a horse as though born on one. The blood lines at the Willow Lane run pure.

It being a working ranch (as opposed to a dude ranch), there was work to be done. I boarded a big gelding named Lightning, whose great girth only occurred to me hours later when my legs began to drift apart like continents. Accustomed to English riding, I was slow warming to western saddles, which seem to weigh more than I do, but I took comfort in the familiar smells — of leather tack and the horse. The feeling I got when I first mounted was also one I knew and loved: you are astride a high creature, feeling small and large at the same time.

Day one. I joined the team. Our task: move fifty or so cattle from one pasture to another as a kind of dry run for the cattle drive; then pluck out a particular cow, no. 9 (identified by a tag in one ear); then drop off in the high hills some toolbox-sized salt licks with the pack horse; then fix a few fences. As I rode over the hill just behind the cedar ranch house and the clutch of Russian willows that gave it its name, I suddenly, and really for the first time, understood the phrase "big sky country." I felt utterly at home. On the range. On a horse.

The cattle moved like an amoeba, an amorphous mass in the middle but when parts darted in one direction the mass was inclined to follow. Cowpokery is about nipping such projections in the bovine bud. Luckily the herd seemed never to move resolutely or quickly, and the horses seemed attuned to the task. This was easy.

Cow no. 9 corrected that impression. While Keith and I fixed fences (actually, Keith worked the pliers, while I took in the glories of hill and sky), the others moved no. 9 down to the corral for an inoculation. Then came the news that no. 9 and her calf had leaped a fence and rejoined the herd.

Our work was just beginning. The wrangler, a shy woman named Cynndae McGowan, kept disappearing into gulches and behind hills at a full gallop in pursuit of no. 9 and calf, both creatures now thoroughly disenchanted and moving both quickly and resolutely. Like two pinto-colored pinballs. Finally, no. 9 was sequestered, thanks more to Cynndae and Keith than to helpers.

Day two. The cattle drive was under way. We followed lonesome roads to a corral, where the cattle were to spend one night before continuing. "What's the best part of being a rancher?" I asked Keith along the way, our horses walking in tandem. As in the movies, there was ample opportunity to ride up alongside a fellow cowpoke and chaw. "I'm doing it," he said with a smile.

My own smiles came harder. By noon I was a wishbone, close to snapping. Another rider offered to swap Lightning for his horse, Gina, and I felt instant relief. I had ridden a zeppelin and was now comfortably mounted on a pencil.

We stopped for sandwiches at an elbow in the road where the cows bunched. Some horses were tethered, some hobbled around us, as we reclined in the high grass. As they grazed and snorted and blew, I thought of the ancient Germanic priests who divined good and bad omens from the neighing and snorting of snow-white horses. That day, the omens seemed as good as the sun was warm.

Gina was a chestnut mare, calm and dutiful. When, later, a gravel truck came on at a dusty clip, the cattle parted like the Red Sea and sought the ditches. I did not, thinking the trucker would slow. He did not. Gina's eyes bulged (much as mine had earlier in the day when I sampled chewing tobacco) and we began a skittish reverse on the road that I worried would send us tumbling backward into the ditch. "It's okay. It's okay," I told her while touching her neck, though it sure as hell was not. Then, suddenly, the accursed truck was gone, taking the twentieth century with it.

The sounds of a cattle drive seemed out of time but oddly familiar, even comforting: the clip-clop of hooves, the lowing of cattle, the idiosyncratic urgings of cowpokes.

"Ssss, ssss, ssss, ssss!"

"Come on, ladies!"

"Hya, hya!"

"Outa there, 43!"

At day's end, LeAnne retrieved us and the horses in a long horse trailer and we retraced roads just traveled. I noted the same patches of brilliant yellow canola; the same ducks, bottoms up, in ponds; the same huge horizon. But everything seemed remote — a topography under glass. The world from a truck was not at all the world from a horse's back.

Day three. The day began at 5:30 A.M. "I can give you a slow but smart horse," Keith had said the night before, after a fine feed of Alberta steak and saskatoon berry pie, "or I can give you a fast but stupid horse. Which do you prefer?" I opted for Mabe: dumb speed.

I liked Mabe. We talked a lot. Mabe would agree, for example, to descend into a damp ditch to move a heifer along, but she would grumble about it, her ears flat against her head and her tail swishing circles to signal her displeasure.

By mid-morning we were in high country, approaching thirty-eight hundred feet above sea level. We ascended a steep rise, and down, far down, in the next valley other cows, other cowboys — Keith's brother's herd — traversed a meadow. They looked like plastic miniatures and the land swallowed all sound of them. Only fence lines, running up and down the vast hills like stitches on a belly, signaled that humankind had ever staked a claim on this wild country.

Keith Lane's forebears had been raising cattle in the Porcupine Hills since 1909. How strange my nostalgia for a life I had only read about. Was it because the cowboy life is so physical and in some ways unchanged from the nineteenth century? At a corral flanked by hills so cragged and steep I wanted to call them mountains, the Lane brothers' herds blended and rested before the final push to the forest reserve.

Beyond that point, roads ended and a new world of lush meadows began. I trailed a cow and her calf as they strayed into grass up to my stirrups. I committed to memory the swish as Mabe cut a swath through the grass; the rising sweet smell of wild mint, with its purple spiked bloom; the brown-eyed susans and the rampant color of all those wildflowers. Contentment washed over me. I loved the quietude and the sweet simplicity of our task: move cattle, at the speed of cattle, to where they must go. It seemed a cleansing thing to be on horseback, neither wholly at work nor entirely at play but in the blessed middle.

That evening, back on the porch of the Willow Lane, we all sat and for hours talked horses. We recalled the moods of our mounts that day, we mused on their intelligence (or lack of it) and we heard horse stories and horse philosophy from Cynndae and Keith — while the horses themselves hobnobbed in the fields beyond the gate, tails spanking flies, heads down to the grass, now and again breaking into a playful canter, manes flying like streamers in the wind.

To ride in near-wild places is to remember a time when individuals lived and died by their partnership with a horse. "The affection between man and mount," wrote the authors of The American Cowboy, "cemented through months of sharing privation and danger and adventure, was therefore an almost human thing; and many a cowboy . . . would gladly share the last of his water canteen rather than see his equine friend go thirsty."

Though I rode her only a few days, I got to like Mabe — her spirit, her cantankerousness, even her impatience to get a move on. But what happens between horse and human when they spend years in each other's company, as Aime Felix Tschiffely did?

The Swiss-born Tschiffely, thirty years old when his trip began in 1925, had been teaching at an English-American school in Argentina for nine years but, as he put it, had fallen "into a groove." (Fallen "into a rut," we would say today.) The notion of a great trek had fired his imagination for years. A photograph of him in his book *Tschiffely's Ride* shows an agreeable, freckled man, his arms crossed, his shirt sleeves rolled up thickly, past the elbows, and on his head a wide-brimmed high hat. If you ever read as a child the *Curious George* series (written during Tschiffely's time), think of George's friend, "the man with the yellow hat." That's the kind of hat Tschiffely wore on his adventure.

He was a rugged traveler and an expert horseman, but he could never have accomplished the journey without his two "old pals" — Mancha ("the stained one," in Spanish) and Gato ("the cat").

Before the trip began Tschiffely had the good fortune to encounter Dr. Emilio Solanet, an authority on, and breeder of, Criollo horses. They were descended from fine Spanish, Arab and Barb stock brought to what is now Buenos Aires in 1535 by the Spanish general Don Pedro Mendoza. When the Indians slaughtered the first Spanish settlers, the horses took to the countryside. Hunted by wild animals and the indigenous peoples, toughened by cold and intense heat, drought and hunger, the Criollos survived — but only the strongest among them. Their feats of endurance fell into legend. The two horses whom Solanet recommended to Tschiffely were only partially broken in, though they were not — at fifteen and sixteen years of age — at all young.

Neither horse was terribly graceful, but each had bright intelligent eyes, a thick neck, sturdy legs and a formidable spirit. Just getting to Dr. Solanet's ranch required the two Criollos to travel more than a thousand miles and to forage on sparse vegetation along the way. Mancha and Gato, the author reports in the wonderfully detailed *Tschiffely's Ride*, were "the wildest of the wild." Bringing them to a stable in the small hours was a monumental task, for streets, houses and cars all terrified them. Tschiffely thought he was doing the horses a favor by offering them the finest alfalfa, barley and oats. They declined those delicacies and supped, instead, on the coarse straw he gave them as bedding.

Mancha, sixteen, was a pinto — red with splashes of white, his face almost entirely so — but pintos were rare in South America and other horses spooked when they saw him. His ears always moving, his eyes always fiery, Mancha had the wariness of a watchdog, and whenever strangers came near he would lift one leg in warning, flatten his ears and stretch his neck threateningly. When Mancha wanted something from Aime, the horse let him know by nickering or neighing, rubbing his forehead against his master or nipping him. He would let no one but Aime saddle or ride him and if anyone else tried to, he would resist by bucking and kicking.

Gato was what Tschiffely called coffee colored; we might say buckskin. The two horses were inseparable, but Gato, at fifteen years, was clearly the junior of the two. He never retaliated when Mancha boxed him. Gato, Tschiffely wrote, was a willing mount, the kind who, "if ridden by a brutal man, would gallop until he dropped dead. His eyes had a childish, dreamy look, seeming to observe everything with wondering surprise."

Tschiffely summed up their characters this way: if the horses could fathom human speech, Gato was the one to tell your troubles to, Mancha was the one you took for a night on the town.

Few horses could have done what they did — cross deserts, navigate dizzying mountain paths, penetrate tropical jungles. They were attacked by vampire bats, gnats and mosquitoes, they endured bitter mountain winds and steamy jungle heat

8.3   A. F. Tschiffely and Mancha greeted in New York, 1928, after a
10,000-mile, two-and-a-half-year journey.

and lived on the most meager fodder. Some of the heart-
breakingly steep inclines they climbed were littered with the
bleached bones of mules and burros who had perished trying
to make the ascent. One desert they crossed in Peru was called
Matacaballo, meaning "horse killer," but it seemed nothing
would stop them.

They carried no water, even across deserts. Tschiffely reasoned that because water is heavy and awkward to carry, they
could travel lighter and faster without it. They drank only
when they chanced across water in a lake or river or village.

The three could have perished on innumerable occasions.
Gato, who had an unerring instinct for quicksand and deadly

mudholes but who was perhaps a little too sure of himself on narrow mountain paths, once lost his footing and was skimming down the mountain toward certain death when a tree brought him to a shuddering halt. Gato had the good sense not to stir, but he neighed pitifully to Mancha while Tschiffely gingerly hiked down the slope and carefully unsaddled the trembling horse. With rope and with help, he was able to rescue the horse, but it was a narrow escape.

If Tschiffely saved Gato's life, both horses returned the favor many times. In one incident, Gato refused to go forward, and even the use of spurs only caused him to rear and snort. Finally, an Indian chanced by and told Tschiffely that he stood on the edge of a very dangerous mudhole. The mystery was how the horse knew, for no such mudholes existed in the horse's home territory.

The three travelers nearly drowned in raging rivers, and the crossing of canyons on narrow suspension bridges, some of them 450 feet long and only four feet wide, took much courage and horse sense.

Aime would walk behind Mancha, holding his tail and talking to calm him. By the time they reached the middle of one sagging and particularly flimsy bridge, it was swaying like a rope under a tightrope walker. Mancha was clever enough to stop until the swaying ceased, before proceeding. Had he lost nerve, bolted or tried to turn back, it would have been the end of them all.

In Ecuador, where the trails were no more than thin lines of muddy porridge, Tschiffely passed the saddest, vilest pack animals he had ever seen. Their sufferings, he said, were best left undescribed, but suffice to say "my conception of hell is to be a pack-animal of the Andes."

Along one such trail, Tschiffely came across an odd, mud-covered shape alongside a mule similarly encased in mud. The shape turned out to be a man who wanted to know what Tschiffely was doing "in these blessed [he meant, I think, God-forsaken] parts." When Tschiffely replied that he was traveling for pleasure, the muddy fellow — a surveyor — thanked Tschiffely for saving his life. The wretched surveyor had considered shooting himself, he said, for being such a fool, "but now, having met one bigger than myself, I shan't do it!"

Though the horses were quite attached to Tschiffely, the vestiges of their wildness never left them. Mancha, by lifting his head and sniffing the air, gave ample warning of other humans in the vicinity. The horses could smell panthers and wild animals a long way off and called pitifully to Tschiffely if he staked them in the jungle far from the hut where he slept. And so all slept in proximity. When in Mexico, Gato had to be sent by train to the city for treatment of an infected knee, the departure was hard on all three. "I had a big lump in my throat," Tschiffely wrote, "when the train disappeared around a curve, for I really believed I would never see my dear Gato again." Mancha and Gato must have had similar thoughts since both horses called out desperately to each other.

The horses took turns as pack animals, and though Tschiffely picked up the odd fellow traveler along the way, for the most part Mancha and Gato remained his only companions. He talked to them a great deal, and they seemed to understand some things. If he said *¿Qué hay?* ("What's up?") they would prick their ears and look nervously about. If he said *puma*, they would sniff the air for lion. They knew that *chuck-chuck* meant food, that *agua* meant water; they would accelerate to his *vamos* and stop at *bueno*.

Tschiffely was later asked countless times if he found the journey in the wilds lonely. Never, he said. "The company of a horse or a dog is a wonderful thing, and with Mancha and Gato I never felt the want of any better." Feeding and looking after them, he said, was a reward in itself, and he never tired of talking to them, nor they, it seemed, of listening.

Their journey ended in New York City, where the mayor welcomed them, and in Washington, D.C., where they were greeted by President Calvin Coolidge. In New York, while Tschiffely was busy with lectures and public appearances, one well-meaning army sergeant attempted to ride Mancha to give him some exercise. The man later said that the "hell-pet" had "gone off like a stick of dynamite."

To complete the journey and get home safely, Tschiffely needed a great deal of luck — right to the end. He gave a last-minute lecture to the National Geographic Society about his extraordinary expedition, and the delay caused him to forgo passage on the ill-fated *Vestris*, whose sinking claimed more than a hundred lives.

By then the two horses and their rider had become true celebrities, and they were featured at an international horse show in New York. They returned to Buenos Aires on the *Pan America*, a ship normally off-limits to livestock. The head of the company declared an exception for Mancha and Gato: in their own way, they traveled first-class and free of charge.

Tschiffely briefly considered someone's suggestion that the horses be put in a public park in Argentina but wisely reconsidered it. The two Criollos got the freedom they deserved and both would live well into their thirties. "As I write these last lines," he wrote in *Tschiffely's Ride*, "I can see them galloping over the rolling plains until they disappear out of sight in the vastness of the pampas . . . Good luck . . . to you, old pals, Mancha and Gato."

Barbara Whittome must possess a little Tschiffelyan wanderlust of her own. In the summer of 1995, she left her home in Suffolk, England, to travel to the tiny village of Alexeikovo, near Volgograd, where she bought four Cossack ponies, intending to breed them with her own Arab stock back home.

In her late forties and a rider since the age of three, she had heard that Cossack ponies were the toughest in the world and decided to test the claim by embarking on a six-month, twenty-five-hundred-mile journey across the plains of Russia. Besides, she added, "I wasn't in a hurry."

In Alexeikovo, she first rode Pompeii, a fifteen-three-hand palomino stallion. "I had a connection with that stallion the

8.4   Barbara Whittome and Pompeii in England after their trek across the
      plains of Russia in 1995.

first time I rode him," she told me. "We charged across the
snow-covered steppes and it was love at first sight. I can't
explain it. It's never happened before." To convey his character, Whittome cited a British fictional series from the 1920s

called *Just William*, about a little imp who bears no malice toward anyone but who gets himself into the worst sort of trouble. Pompeii, Whittome said, was just like William. He was not fit when she purchased him, but the journey and the years since have improved his appearance mightily. He still loves long rides, and has lost none of his Williamness.

Whittome bought Pompeii and three other ponies "for peanuts," though far above the going local price, and headed west, but she soon discarded one pony given to biting and kicking. She took turns riding Pompeii, a black mare named Masha and a gray mare named Malishka. Whittome began the journey with Cossack guides, reputed to be tough horsemen, but they whined so much about saddle sores and the rugged pace that she left them behind and sometimes used a compass to guide her, instead. For most of the way, a companion followed behind in a truck bearing oats (the horses went through fifty sacks) and camping equipment.

And while the ponies themselves were relatively cheap to purchase, the expedition cost close to $90,000 and forced her (now former) husband, Giles, to sell off his antique-gun and classic-car collections.

Along the way a friend named Alison Lea, then later, Whittome's daughter Katie Farmer, joined her. There were some terrifying moments. Russian thugs on motorbikes tried to rob their camp, but the mounted women fought them off, and Pompeii, with Barbara up, even trampled one would-be robber. Giles later scored it, "Girls on horseback 1, Russian thugs on motorbike 0."

Seeing a woman on a cross-country trek brought out the best in the Russian people. "They were so warm," said Whittome. "Once I told them what we were doing they nearly collapsed in amazement." The police stopped them almost daily, but when told of the circumstances, stunned officers never even examined their passports. The weather was fierce in the extreme — from ninety-five-degree Fahrenheit (thirty-five Celsius) temperatures and a fortnight of rain in southern

Russia to sleet that came at them horizontally in wintry Poland.

Russian Ride, Whittome's book, captures the frustrations of the adventure — maddening Russian bureaucrats, inept guides, soaring costs, sniping among traveling companions. "I am probably running on boiling rage," she writes at one point in her diary. But the book, like Tschiffely's, also touches on some of the rewards of a long trek on the back of a horse. The joy of waking to the sound of horses grazing a few feet from your tent. The generosity of strangers, especially the horse-loving kind. The starry nights, the dappled shade in the birch forests, the beauty of the dawn, the fenceless lands that redoubled the sense of freedom and journey.

And like Tschiffely, Whittome came to know the horses very well. "The horses' characters are beginning to emerge," she writes in *Russian Ride* on day thirty-nine. "Pompeii is still the rather dreamy stallion I fell in love with last year while Malishka (Muffin) is extremely knowing, and very much aware of everything going on around her. She is also a dreadful coward. Masha has the typical Russian pessimistic attitude to everything, and ought really to be rechristened Eeyore."

A photograph taken along the way shows a smiling woman on horseback wearing a Russian fur hat. The Cossack ponies would prove their legendary hardiness, though the reputation of Cossack riders took a beating. Whittome is a natural traveler: keenly multilingual, immune to homesickness, appreciative of creature comforts but never their prisoner.

"It has been the experience of a lifetime," she said later. "At times it has been incredibly frustrating, but if I had the opportunity to do it again I would like a shot. The thing that sticks out in my mind is the freedom. You just roam wherever you please."

In 1998, Whittome, now employed by Lloyd's of London (perhaps they'll insure her?), plans next to ride in the footsteps of Carpine, the portly, elderly monk sent by Pope Innocent

IV on a diplomatic mission to Genghis Khan. This in 1245, when a sixty-three-year-old man was indeed aged, and well past a two-year, ten-thousand-mile journey from Lyon in France to Lake Baikal north of Mongolia. Easy to see, then, who Whittome's heroes are.

In the meantime, she has sold the Cossack mares to endurance riders "who would continue to prove the point I think I have already made." But she still has Pompeii and the handsome foal he sired with Masha, called Ashibka — the name means "mistake" in Russian. But make no mistake. There will be other journeys on horseback for Barbara Whittome.

Another British woman, Christina Dodwell, has written widely about her own journeys on horseback through Africa (five thousand miles), Papua New Guinea (eight hundred miles) and Iran and Turkey (fifteen hundred miles).

Dodwell is a British broadcaster and author who details some of her adventures in *A Traveller on Horseback*. One of her ancestors was captured by the French in what is now Iran and eastern Turkey during the Napoleonic Wars. He was given his freedom on condition that he not return to England, and so he traveled the region — by mule. Christina Dodwell aimed to go over some of that same ground — alone, on horseback and at a time when the Ayatollah's revolutionary Iran was at war with Iraq.

The poor timing meant she was arrested five times, and while she managed to talk her way out of jail, she did spend as long as twenty-one hours incarcerated. A veteran traveler, she kept her wits, using dignity and courtesy and forming alliances as the need arose. During a previous journey in New Guinea, several youths attempted to rob her of her saddlebags but let her bargain for the contents. During the banter that followed, she garnered the name of their village and thus led the boys to fear she might report them to their elders. She kept *most* of what they had taken.

Part of Dodwell's mission in Iran was to research Kurdish

Arab horses. Near a Turkoman village she met a remarkable woman named Louise Firouz, who had made it her life's work to reestablish a breed of miniature Caspian horse, extinct for more than a thousand years. A relic of the ancient Persian Empire, the horse can be seen on bas-relief sculptures amid the ruins of the old capital of Persepolis. Firouz had noticed, among wild horses roaming the region, occasional throwbacks to the breed. She had collected numbers of them, built up a herd and exported some to Britain and the United States.

Dodwell rented and borrowed horses along the way, but in eastern Turkey, she purchased an iron-gray Arab stallion called Keyif, which means "high-spirited." The horse, who ran more than he walked, would take her through the mountains and to Mount Ararat, reputed to be the resting place of Noah's Ark. Over the course of several months, Dodwell got "delightfully lost," dressed as a man to discourage would-be bandits, enjoyed the warmth of peasant hospitality and chanced upon "stunning Christian ruins, lost and neglected, their crumbling domes soaring to heaven." At the end of her journey, Dodwell sold Keyif to a forest ranger who happened both to need a horse and to be enamored of horses. On a rattletrap bus to Istanbul and thence home, Dodwell knew she would soon be glancing through an atlas and pondering the next epic ride.

Were there to be such a thing as a long-distance riders' hall of fame, the membership committee might consider the following for inclusion. Whatever inspired them — duty, fame, daring — they redefine the hard ride.

Rafael Amador might merit a place in the patriots' section. In 1834, General Santa Anna of Mexico learned of a plot to seize and sell the California mission lands. It was imperative that a message of warning reach California. He chose as his courier Rafael Amador, who set out on the morning of July 25, 1834. Between Mexico City and Monterey lay more than twenty-five-hundred miles of mountains, jungle, rivers, deserts and hostile Indians. In Mexico he galloped the whole way, stopping only to change horses. Through the Apache territory he traveled exclusively at night. In Colorado he was ambushed by Indians and escaped only with his clothes and the precious message from his general. He was then forced to walk 150 miles through terrible heat. Finally, he arrived, only forty-one days after leaving Mexico City.

Slim John Brown might also win a place among the bravest of epic riders. Old westerns often set this scene: the cavalry are surrounded by Indians, and it falls to someone — the man who draws the short straw — to make a dash through the cordon and bring help. But in Brown's case the circumstances were frighteningly real.

Brown had in his possession a number of cigarette papers, each stamped with the seal of his superior officer and bearing the words "Believe the bearer." Brown had horses shot out from under him, procured fresh ones using the cigarette papers and rode one horse to death. His wild ride from Los Angeles to San Francisco, six hundred miles, took him four days. And the cavalry did indeed come to the rescue of their fellow soldiers.

Then there is Felix X. Aubrey, whom one western historian calls the supreme rider of the West. Perhaps only a high-stakes gambler could understand why he almost sacrificed his life to win bets. In 1848, this jockey-sized French Canadian galloped the more than 780 miles between Santa Fe, New Mexico, and Independence, Missouri, in twelve days. The locals were astounded. Aubrey then bragged he could cover the same distance in eight days. He did, but he rode three horses to death, walked or ran forty miles, slept only four hours and ate nothing for three days.

Finally, Aubrey accepted another bet, this one for $1,000, that he could do the trip in six days — if he used a relay system (as the Pony Express would twelve years later). This time he reportedly did not stop to eat, sleep or drink, and strapped himself into the saddle. One mare, a dun called Dolly, took him well over 180 miles after an alternate horse fell prey to

Indians. Aubrey arrived in Independence, according to one account, "a ghost of a man who could speak only in a hoarse whisper when lifted from his blood-soaked saddle." Still, Aubrey was right on time and right on the money.

If Aubrey was the hare, John Wesley was the tortoise, slow and steady. Perhaps no rider will break the record of this eighteenth-century English evangelist. Prints from that time show him in the saddle, with loose reins, and the rider completely absorbed in his Bible. "Though I am always in haste," he once wrote, "I am never in a hurry." During his lifetime as an itinerant preacher, he rode 288,000 miles. Consider, too, Len Crow, the mason and "Christian cowboy" from Barrie, Ontario, who rode from Fairbanks, Alaska, to El Paso, Texas, (4,200 miles) in 1996 to raise money for his church's work in the Phillipines. His mount most days was an Arab gelding called General.

Without doubt, there should also be a place in the hall of fame — among the adventurers — for Fred Burnaby, and especially for his horse. Some long-distance riders have covered the distance more quickly, and no doubt some might have done it more gracefully. But what sticks in my mind about this thousand-mile round trip through central Asia in the dead of winter in 1876 is the image of a very large man on a very small black horse and the pluck of both.

Burnaby, captain of the Royal Horse Guards, weighed about 224 pounds. The son of a clergyman, he was born with what his nanny coined a "contradictorious" nature. He seemed to possess a healthy contempt for authority, represented by his father's church, and he joined the cavalry at the age of seventeen. The world was indeed his oyster, and for fifteen years he served in Central and South America, Egypt, Spain, Russia and Morocco.

On leave in the Egyptian city of Khartoum in 1875, he spotted a notice in a British newspaper that the government in St. Petersburg had declared a ban on all foreigners' travel in Russian Asia. An Englishman had recently been expelled for defying the decree. Captain Burnaby had long wanted to go there, and now that it was forbidden, it seemed more enticing than ever.

He set as his goal the ancient town of Khiva, in what is now the republic of Uzbekistan. Worse than the ban on travel was the forbidding area to be traversed — a vast snow- and salt-covered track. "The cold of the Kirghiz Desert," Burnaby wrote in *A Ride to Khiva* in 1877, "is a thing unknown I believe in any other part of the world, or even in the Arctic regions. It blows on uninterruptedly . . . [like] the application of the edge of a razor."

In Kasala, after he let it be known he needed a horse, a series of sorry mounts were paraded before him. "Their ribs in many instances almost protruded through the skin . . . Except for their excessive leanness they looked more like huge Newfoundland dogs."

Burnaby typically rode large and powerful horses to convey his bulk, but none existed in Kasala. In the end he selected a fourteen-hand black horse with a thick coat and a rather elaborate gilded saddle. When Burnaby and his retinue set out on January 12, he wore several layers of sheepskin and other clothes and further weighted the pony with huge iron stirrups. The little horse was heard to groan as the full 280-pound weight settled onto his insubstantial back. A Tartar servant on a camel carrying food packs confidently predicted that Burnaby's horse would soon collapse under the great weight and they would all make a meal of him.

The little horse, though, plodded on, into blinding gales and over frozen rivers, through mountain passes and across great plains. After one morning trek of seventeen miles, at times through snow close to three feet deep, Burnaby marveled at the endurance of the horses. "The one I rode, which in England would not have been considered able to carry my boots, was as fresh as possible after his march." Three hundred miles and thirteen days later the horse was thin but still full of vigor. Burnaby's reward at journey's end was that the khan, or ruler, of the city of Khiva did indeed welcome him, and after a nine-day rest the little black horse turned back for home stronger than ever. Over that forbidding terrain, Burnaby and his horse averaged forty miles a day and even galloped the last seventeen miles.

The fate of the black horse, never mind his name, was not recorded. We know only that after buying him for five pounds Burnaby sold him for just over three.

Worthy of a place among the marathoners, too, are all those who traveled by covered wagon across North America. Classic chronicles of their journeys, such as Francis Parkman's *The Oregon Trail*, still make for chilling reading today. "It's a serious thing to be travelling through this cursed wilderness," one man says to another around a campfire in 1846 as all contemplate the fate of four men gone missing and last seen being dogged by Indians "crawling like wolves along the ridges of the hills." The horse-drawn wagons, which Hollywood would help make a symbol for quest and journey, moved at a snail's pace through the terrible beauty of the plains. The toll on both horse and human was terrific.

Parkman refers to the "ragamuffin cavalcade" that followed the banks of the Upper Arkansas River: "Of the large and fine horses with which we had left the frontier in the spring, not one remained: we had supplied their place with the rough breed of the prairie, as hardy as mules and almost as ugly," yet even they grew foot-sore. Worn saddles, rusty guns, buckskin-clad riders — they looked a sorry sight.

"The cursed wilderness," on the other hand, was undeniably breathtaking. Parkman describes the "undulating ocean" of grass, its seed "as sweet and nutritious as oats" and impossible for their hungry horses to resist, its fronds almost as tall as the horses themselves. And when the sun set, the western sky was left "all in a glow."

Centuries before the long line of wagons inched west, the Gypsies had begun their diaspora from northwestern India into Europe. The Rom ("the people," in Romany, the Gypsy

8.5   Gypsy talking to grey horse in Romania: the gypsy diaspora was horse-powered and horse-centered.

language), traditionally lived in *tsara*, covered wagons pulled by horses or donkeys. Nomads all their lives, they worked as tinsmiths, horse dealers, chimney sweepers, peddlers, scavengers, basket makers, farm laborers, fortunetellers and musicians. "Tinkers," the Irish called them, and a "tinker's dam" (as in "I couldn't give a . . .") originally referred to a ring of dough placed around the leak in a pot when the Gypsy tinsmith was using solder to repair it.

Burdened with reputations as thieves and connivers, the Gypsies suffered horrible persecution (five hundred thousand died in Nazi gas chambers), which only made the open road more attractive. In the United Kingdom, Gypsies are known as "Travellers." "We have to travel. It's in our blood," a Gypsy once said. "I wouldn't mind a house," said another, "as long as it had wheels on it." For many centuries, the image of the Gypsy camp has centered on a clutter of caravans at the side of the road, washing hung on trees near an open fire and a horse or two nibbling on grass.

The Gypsy connection with horses is well-known among

horse people. Nicholas Evans came by his idea for *The Horse Whisperer* during a trip to the Dartmoor region of England, where a blacksmith told him about Gypsies who talk to horses.

The best long-distance horse is sometimes not a horse at all but a pony. In 1897, a twenty-four-hundred-mile endurance race between Sheridan, Wyoming, and Galena, Illinois was won by two brothers who simply caught and broke two mustang ponies running wild on the plains. During the arduous ninety-one-day trek, competing horses were allowed no grain and were to forage along the way. "Praise the tall, but saddle the small" is an old Mexican saying: the two winning horses, as fresh at the end as at the beginning, weighed 750 and 900 pounds.

Then again, sometimes neither horse nor pony wins the marathon ride. To mark 1976, the bicentennial year in the U.S., the Great American Horse Race started in Sacramento, California, and ended in New York. Each entry, and there were one hundred, was allowed a remuda of two — one horse to ride and one to follow. The race began the last weekend in May and ended Labor Day in September. The winner was a man named Verl Norton — riding a mule!

CHAPTER 9

MY KINGDOM
FOR A — PONY

I never felt such power and action in so small a compass.
A BRITISH RIDER, WHO WEIGHED 196 POUNDS,
PRAISING THE TWELVE-THREE-HAND EXMOOR PONY
HE RODE TO THE HUNT IN 1820

HER FARM AT Plettenberg Bay, some five hours east of Cape Town, affords a compelling view of the fertile valleys where the pink heather grows as high as the sixteen-hand gelding who would figure in this tale. As impassioned a horsewoman as I have ever met, Vickie Rowlands runs nine-day riding treks over a three-thousand-acre territory. The photographs she showed me left me enticed; her descriptions of the rides seemed ethereal: "I like to sleep outside under the stars . . . it's quite safe . . . we have leopards and snakes, but they won't bother you, and with the horses you can get close to the wildlife . . . you can hear the sea up there." But one morning in 1990, she told me over coffee during a recent visit to Canada, her little patch of paradise turned rather hellish.

Vickie awoke at dawn to the chilling and unmistakable sound of equine assault. A former national distance-riding champion in South Africa, a horse breeder and lifelong trainer, she knows all the sounds — sweet and otherwise — that emanate from horses. The one coming from the near paddock filled her with dread. She screamed. With her husband, Kevin, she ran outside, the two of them ludicrous in their underwear, both alert from the adrenaline pumping in their veins.

A small fourteen-hand mare had thrown a foal ten days beforehand, and mother and baby had been separated from the twenty-seven other horses and kept in a field by the house. But in that field an awful drama was unfolding, and by the look of all involved had been under way some time.

A big gelding had jumped the fence, intent on dispatching the foal. The mare, her sides heaving, her flanks in a lather, was too exhausted to help any more and weakly whinnied to her desperate foal. The screaming foal fled the gelding as best she could, darting and doubling back, but she was losing her thin hold on life. Vickie and Kevin looked on, helpless. At one point, though, the little foal lost her footing in the melee and slipped under the fence, like a runner from third who slides past home plate. She ended up on her side in the other field, where she seemed at last safe from the gelding. "Thank God," Vicki uttered.

But the wide-eyed gelding, silent through it all, was not done yet. He leaped the fence, the other way this time, picked up the foal by her neck and shook her — "as a dog would a rat," Vickie said. The gelding had one thing in mind. *Kill the foal.* Kevin picked up a fence rail as thick as a fist and pounded the gelding over the head. At this the attacker relented and the foal fell hard to the ground, but the gelding again bent low and took the leggy little horse in his jaws, and once more gave her that savage shaking. Kevin cuffed the gelding with renewed vigor, this time breaking the fence post over the horse's head.

Throughout, the other horses had maintained their positions two hundred yards away at the far end of the adjacent field, minding their grazing business as horses are wont to do. But then came Hayla, Vickie's daughter's pony, the thirteen-two-hand offspring of an Arab sire and a Welsh pony mare. Ears pinned back, she had seen enough. Hayla put herself between the gelding and the foal and drove the foal at a gallop, their sides so close they touched, into another field. Then she nuzzled the petrified foal and laid her head on the wee horse's back as if to say, "There, there. It's over now." The rescue had bought enough time for Kevin's pole to give the gelding pause and for Vickie to close a gate. It was indeed over.

The foal did survive the attack: the hideous swelling on either side of her neck where the gelding had closed his teeth eventually subsided, though dents would remain to remind Vickie of the day the gelding declared such uncharacteristic yet deadly purpose. As for Hayla, Vickie told me, "she was, funnily enough, not the nicest pony in the stable." Like some of her kind, she was a little mean tempered, but if there is such a thing as pony heaven Hayla has surely earned her place in it.

What to make of the episode? Vickie said she had seen other incidents of altruism in the wild. To the film footage she once saw of a hippo rescuing a springbok from an alligator she could now add the story of the Arab-Welsh pony who came to the aid of a beleaguered, terrorized foal.

For many of us our first moment in the saddle — rocking horse and carousel aside — is a rent-a-pony at the county fair. A dollar a ride: beaming toddler grips pommel, proud parent grips toddler, bored handler leads bored pony in circles. All very tame, but for the young rider, all very heady. The lineup for such rides is always long.

A clever cartoon once turned upside down the cliché of the young girl begging for a pony of her own: mare pony addressing her foal, says, "Yes, I know you'd love a little girl of your own, but where would we keep her?"

9.1   A young girl's obsession with horses often begins with ponies.

At first blush, love for the horse is actually love for the pony. What I have come to feel, after reading about ponies and after talking to pony-handlers, is a deep respect for these often maligned and much misunderstood creatures. The sense of surprise in the epigram to this chapter — "I never felt such power and action in so small a compass" — mirrors my own. The burly man who said those words expressed astonishment that the little Exmoor pony he rode was able to jump a fence, with him aboard, when the fence was eight inches taller than the pony. Man and pony then galloped the eighty-six miles from Bristol to South Molton quicker than the fastest stage-coach of the day.

The Romans rode Exmoors when they ruled what is now Britain. The Bayeux Tapestry shows William the Conqueror riding an Exmoor when he landed in Britain in 1066. Elwyn Hartley Edwards, the British equestrian authority, praises the Exmoor pony's independent nature, adding that when properly schooled the pony makes a brilliant mount for children.

The best competition horse, he muses, might be a pony-horse cross (Exmoor-Thoroughbred), one that offers a strong constitution, hardiness and intelligence and "that peculiar sagacity that contributes to the 'streetwise' quality of the pony breeds."

Throughout history, ponies have been ridden to the hunt, to flocks, to market and to war. Did our ancestors ride ponies because horses were not available — or did they actually prefer ponies? And when is a pony no longer a pony but a horse? The distinction is a tricky one. Polo ponies and cow ponies are typically horses but may also be ponies. Many North American experts say that a pony is typically under fourteen hands two inches.

Edwards sets the dividing line at fifteen hands but insists that proportion, not height, marks the real difference between pony and horse. Largely because of the horse's long legs, he says, the distance from withers to ground exceeds the length of the horse's body. The pony, shorter in the leg, is longer from head to tail than from withers to ground.

Some pony lovers and fanciers maintain it is quite easy to distinguish pony from horse: the pony, they say, is smarter and inch for inch can far outjump a horse.

An ocean away from Vickie Rowlands and her cherished Hayla, another woman, Adele Rockwell, heaps praise on the pony, the Welsh pony in particular. The Rockwellian defense of ponies in general and that breed especially — against all who call the pony a lesser cousin to the horse — is spirited, affectionate and unyielding.

On her pony farm north of Toronto, Ontario, Rockwell talked with me about the longing to ride, citing a proverb that may have its origins in a time when the poor walked and the rich rode. "If wishes were horses," Rockwell said, "beggars might ride."

J. Frank Dobie, an historian of the old West, once turned the phrase on its head to convey the plenitude of wild horses in South America two centuries ago: "Horses were as cheap as wishes: beggars rode." The original phrase comes from a seventeenth-century gatherer of proverbs named John Ray. Most of us have forgotten his name and his book, *English Proverbs*, but perhaps thanks to him we still have "Misery loves company," "Blood is thicker than water," "Money begets money" and, of course, "If wishes were horses . . ."

The publisher of several Canadian horse magazines calls Adele Rockwell "one of the genuine treasures in the pony world." I took her to mean that there is the real world, there is the horse world and then there is the pony world.

For more than forty years Adele and her husband, Dick, have bred and trained Welsh ponies at their farm near King City, a grand name for a small place. The Southern Ontario countryside rolls a little crazily up there, like a giant loaf of egg bread, full of creases and valleys but perfectly rounded on the hilltops. On the day we spoke a few dozen ponies, gray most of them, were gathered on the crest of one hill where a young woman was putting out hay.

Dick greeted me at the door of the farmhouse, or at least,

9.2   Dick Rockwell and Ardmore Gangway: seeing the world through the pony's eyes.

let me in on his way out to the ponies. He is a man in his mid-seventies with a goatee, bushy black eyebrows and a voice that reminds me of water moving over a gravel bottom. The house is like many a horse owner's house — a little plain, a little cluttered, a house that does the job but is likely not as well appointed or organized as, say, the green barn (home to seventy ponies) where Dick was headed.

Adele led me into the front room, which, it turned out, was not a room at all but a miniature museum. Shelves were laden with sculptures of pony heads and pony bodies, photographs of ponies alone and with their proud handlers, paintings and sketches of ponies and more ponies, silver trophies and copper plaques — so many that to pluck one (as I asked Adele to) risked a shelfslide among the rest.

The two north-facing windows in the room let in only a feeble light that gray day, so the several table lamps had been switched on. Reflecting the room's abiding color — the pink and red of hundreds of horse show ribbons — the lamps cast a warm, liquid light.

The room was the color of fair skin too long in the sun: ribbons lined every wall where it met the ceiling; great racks of them had been set up on wallboard leaning against chairs and hassocks. The room was effectively wallpapered with ribbons — this one from a pony show in Devon, Pennsylvania, thirty-eight years earlier, that one from last year's Royal Winter Fair in Toronto. Ribbons covered the hearth and even the lampshades wore garlands of them.

How could anyone sit where I did and *not* talk ponies?

Rockwell sat by a north window, gray sweater buttoned at the front and ending at the neck in a red kerchief; gray slacks; gray running shoes — clothes as utilitarian as the house. Sometimes she would look to the floor or the ceiling — ribbon-free zones — for the answers to my questions.

Of her childhood in Toronto in the 1920s Adele recalled, "I could draw a horse before I could write." Adele Davies liked to go out to her father's Thorncliffe Stable and draw a horse called South Shore, who would win the 1922 King's Plate — the actual football-sized silver trophy was up on the shelf in the pink room. The Queen's (or King's) Plate, the longest running stakes race on the continent, dates from 1860; the first Kentucky Derby came later, in 1875.

Rockwell's father managed the Don Valley Paper Company (now known as Domtar) and her grandfather owned the old Thorncliffe Race Track and farmland in what is now urban Toronto. It was there that he and his six horse-mad sons raised Thoroughbreds.

Look in *The Plate: A Royal Tradition*, the flashy history of the Queen's Plate, published in 1984, and you will find the Davies family well represented. There's Adele's Uncle George in his tux — top hat in one hand, trophy in the other — as he offers a muted smile to the camera after South Shore's victory. The horse was unique in Queen's Plate history: she was one of only a very few mares to win the plate and the only one to produce offspring who did the same — the colt Shorelint in 1929 and the filly Sally Fuller in 1935, Thorncliffe Stable horses both.

The Davies family's connection to horses goes back to the nineteenth century and Adele's grandfather, Robert Davies. The author Louis Cauz, in *The Plate*, calls him "one of Canada's most respected horsemen." A breeder and twice winner of the Queen's Plate, he personally rode a horse called Floss to victory in the 1871 running of that race.

The family colors had always been canary yellow with black stripes and Shorelint's jockey, Jaydee Mooney, wore them in the 1929 victory. Mooney also rode Black Gold to victory in the 1924 Kentucky Derby, inspiring Marguerite Henry's novel.

"I was bred into the horse world," said Rockwell. E. P. Taylor, the legendary builder of Windfields Farm, where Northern Dancer was born, also felt a genuine passion for horses and bought his first riding horse, Blue Wave, from Adele Rockwell. Although she would breed and school horses, her first love was, and remains, ponies, specifically the Welsh ponies — which makes sense, given Adele's very Welsh maiden name of Davies.

At one point in our conversation, she disappeared into a back room to fetch one of her many fine paintings (she is also an accomplished sculptor) and Dick had come back in from the barn. Both eyed the rendering of a pony called Delphi, and offered it as proof of "pony character." It was there in the face, in the eyes, as clear as day.

"Do you see it?" each asked in turn.

Lips pressed, eyes set in a muscular squint, I dutifully considered the painting. Saw the ears pointed attentively. Saw a brightness in the eyes. Saw . . . well, a pony. Just a pony, really. I felt like a student who plays cymbals in the high-school band (which, in fact, I did) being asked by two conductors to read a sheet of music from a symphony. I could pick out the percussion notes but not much more.

So I countered their question with one of my own.

"What is it that you see — even in a photograph, never mind a painting — that I don't?"

"If you can look at a horse and judge character," said Adele, "you have accomplished a lot. Many people look at the legs first. But when I meet a person, I look to the face, to the head. It's the same with a pony." To which Dick added, "When we speak of a pony, and whether it's a good one, we use the phrase 'pony character.' This is unique to ponies and it's difficult to explain. We think of ponies as living persons in their interactions with other ponies and with people."

What the Rockwells saw in the painting of Delphi, what they saw with absolute clarity, was the pony's temperament — a kindness in the eyes, attributes of sweetness, gentleness and friendliness and an overall impression of quality.

The Ultimate Horse Book refers to the Welsh pony's "quality riding action, adequate bone and substance, hardiness and constitution and pony character." However, the author, Elwyn

Hartley Edwards, another fine Welsh name, does not define "pony character."

Though both Dick and Adele tried hard that day and later on the telephone to make me understand pony character, I fear it eludes me yet. I understand that horse and pony, though members of the same species, are not members of the same family, and that while most ponies have pony character, no horse does. Pony character, Dick maintained, is something in the pony's look and disposition, something passed on in breeding, *a* quality but not necessarily *good* quality. I imagine that a pony with pony character has a strong sense of his own self — Dick brightened appreciably and laughed when I told him that. But he knew, and I knew, that my education in ponies had only begun.

Why is it, I asked him, that in many stables ponies win rotten reputations — mischievous, cantankerous, Napoleonic little biters who sometimes boss horses twice their size?

Dick had heard the complaint before, and he worked up a pretty good head of steam as he tackled it. "Ponies have long been viewed as second-class citizens. I was reading something in a magazine the other day — 'She has now outgrown ponies and graduated to horses.' That's wrong. Many people have had a great deal of pleasure and success with ponies. When they can't ride them any more because they've gotten too big, they spend five or six years trying to find a horse as good as the pony was. Lots never find one." Dick argues (with more passion than proof) that ponies, pound for pound, hand for hand, can far outjump a horse. He cited the case of an eleven-hand yearling colt who could jump a four-foot fence: "That's not at all unusual."

Seeing the world, as they often do, through ponies' eyes, has led the Rockwells to two conclusions: one, ponies are smart (which sometimes makes them enemies); two, humans are not always smart (or at least smart enough to realize that the pony is not the problem).

Witness the conundrum of how to introduce a pony to saddle: the tiny rider has too little experience; the experienced trainer, too much bulk. "The result," says Dick, "is that many ponies are not sufficiently schooled. Because of his intelligence and disposition, the pony is generally willing to do what is asked of him. But if the trainer is not capable and the pony lacks confidence in him, then the trouble starts."

Diane L. Huber, a veteran teacher in U.S. Pony Clubs and now living near Ithaca, New York, echoes those sentiments. "Ponies are honest," she told me, "they have a lot of integrity, but even a good pony in the wrong hands . . ." She offers a story to illustrate her point.

Huber had loaned her own pony (by a Standardbred-Shetland stallion out of a Welsh mare) to a young rider for a summer. The pony, called Tinsel Time, was highly trained, fearless and dependable, a ten-year veteran of Pony Club rallies. "I think the absolute world of her," said Huber, who has spent a lifetime around horses and ponies. But that summer, the unsupervised little rider jumped the pony willy-nilly and got her so flustered she began to stop at fences. "Unheard of," said Huber, who eventually set the pony straight again.

"There must be a million ponies out there in the same circumstances or worse. A little kid on a nice pony can do a lot, especially with good, decent guidance. You'll see a kid on a pony, and he's frustrated, and so is the pony — it works both ways. Then all of a sudden, the harmony comes. What's true on a horse is true on a pony — every minute is different. There's something always going on."

Adele Rockwell's father bought her first pony, a Shetland, when she was two. She painted an arresting and nostalgic picture of a time seventy or so years ago when she and her brother rode ponies in the middle of a city and felt a freedom and a carefreeness that may now be gone forever — at least from the urban landscape.

"My brother and I used to ride our ponies in the Don Valley," part of a heavily treed greenbelt that still winds its way through the city of Toronto. "I must have been five or six, and Bob was seven or eight. We were not supposed to leave the forty-two-acre property, but Mom and Dad would have gone to the races in Hamilton or Fort Erie and we'd have been babysat by an aunt. We'd go off and ride the ponies — we'd be gone for the whole day!" Their destination was a certain tree in the middle of the Leaside Flying Field and they would spend hours there watching the biplanes, and later the monoplanes, land and take off.

"We had to cross railway tracks to get there," she continued. "The ponies would step on the ties and make their way across. We were so young, but we were in the hands of the ponies. The ponies would look after us."

By 1935, Rockwell had become a competent rider and had learned from her father how to judge quality in a pony. She went to the Royal Winter Fair that year and came back astonished by the dressage horses she had seen. A Captain Tuttle, an aspiring Olympian and American cavalry officer, had shown what balance and sensitivity his horses were capable of. One called Si Murray could canter backward; another called Vast he reined by a thread.

Like many people born to ride, Adele Rockwell has a photographic memory for the ponies and horses in her life: Creta, the eleven-two-hand Welsh pony, a bay: Pico, a pinto pony from the rodeo; Cellophane, her first horse.

In time she met and married Dick, who was as passionate about horses as she was. They bought a farm in Toronto not far from the original Windfields Farm (both farms were long ago carved up and paved over). The choice was between having horses and having a car: Ford lost out to the feed store.

Dick and Adele became active in the Eglinton Hunt Club and then the Pony Club. The latter was a British invention designed to teach children riding skills despite the onslaught of mechanization after the First World War. Like the Boy Scout movement, another British idea, it spread all over the world.

A Canadian colonel launched its counterpart in Canada in 1934, then passed the reins on to Adele in 1939 when he went off to war. She became district commissioner of the Pony Club of Canada and held that position for ten years. Membership then cost 75 cents a year, and a Pony Club button set you back another seventy-five pennies. Most children brought their own ponies, but some just hung around, drawn to the ponies like ants to honey.

One such boy was named Brian Murray. "I'd lose sleep over that boy," Adele recalls. "During the war the Pony Club put on a horse show in Riverdale Park, a fund-raiser for the war effort. He had no mount, but he was so keen. He'd sit on the fence, pat my friend's pony, hold him if asked. His father was blind, his mother took in washing but he had a paper route. And that was the only reason he was able to afford the $1.50 and join the Pony Club. Eventually, the Pony Club moved to new headquarters, but the bus would only go so far and Brian would have to walk at least a mile and a half. In winter we'd hold meetings at someone's house. Each kid paid $.25 toward cocoa and cookies, and we'd have speakers — vets, blacksmiths — lectures on stable management, films. That boy just soaked it all up. But was it right, I thought to myself, to encourage his interest in horses?"

Years later Adele got her answer in the mail. Someone sent her a clipping from the *Racing Form* — a story from the U.S. about a leading trainer. Brian Murray. He had started by walking "hots" (cooling down horses after workouts) at Woodbine Race Track in Toronto and bandaging horses. An owner was going to put down a crippled horse, but young Murray nursed him back to health and won, said Rockwell, "more money than he had ever seen in his life." The *Racing Form* article confirmed that his dream of a life with horses had been fulfilled; the dream had begun with ponies.

The Pony Club, for Adele and for all her little charges, was a way of life. Jim Elder, Norman Elder and Tom Gayford — Olympic or Pan American Games champions all — took instruction as boys from Dick and Adele. Likewise, in the United States the Pony Club has long been a launching pad for champions. The eventer Bruce Davidson, dressage rider Robert Dover, show jumpers Tim Grubb and Michael Matz all cut their teeth in the Pony Club.

The pony taught them their riding manners, and clearly taught them well. Like the Rockwells, Jim Elder was born into the horse world. A pony named Madelaine, a skewbald Shetland-Welsh cross twelve hands high, introduced the boy to walk, trot and canter.

Elder remembers the pony and the woman who taught him to ride that pony. His farm is only a few minutes' drive from the Rockwell farm. Perched on the edge of his couch before a great fieldstone hearth, Elder ranged over his life story, moving back and forth in time. Still trim, he told some stories to make a point, others to make me laugh, some because he could not resist. His is a life peppered with horses and when this man talks horses the words cannot come fast enough.

"The Don Valley was our playground then. It was a wonderful valley with dense woods. This would have been the late 1940s and 1950s. Those were the Pony Club days, and Adele would give us lessons. She just loved the sport and passed it on. She and Dick became the grand old mom and dad of ponies. We'd have pony shows and Adele would give each of us an opportunity to run the show and do each job. My strongest friends are from those days, and most of us are still riding or still involved with horses."

When Elder got too big for Madelaine, he went on to ride horses and fashioned a sometimes glorious and ofttimes challenging life in the saddle. He rode in the hunt, played polo and helped set up the Toronto Polo Club. His souvenirs of a life in the saddle include two broken backs (one from steeplechasing and one from schooling a horse) and several Olympic medals. My sense is that he rode out of the same gleefulness he took to riding his twelve-hand pony, a pure delight he still seems to possess.

By the late 1950s, the Rockwells were raising Welsh

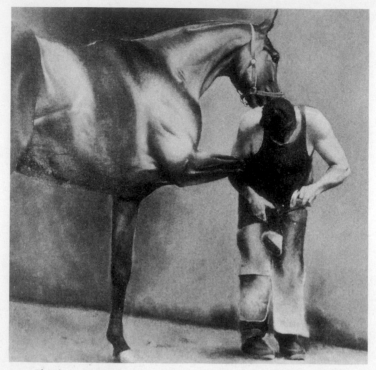

9.3　If polo ponies and cow ponies can be this imposing, when is a pony no longer a pony but a horse?

ponies. Too many Pony Club kids were riding unsuitable and badly trained mounts.

The first pony they entered in the Welsh class, a new class at the 1958 Royal Winter Fair in Toronto, was a pony named Ardmore Gretton Sunlight, from the newly registered Ardmore Stud Farm of Adele and Dick Rockwell. "She was supreme champion," Dick recalled, "and she showed against all the mares and all the stallions. They pinned this three-foot-high ribbon on her — she had to lean to one side to avoid stepping on it. When we got her home she came down the ramp of the horse van and surveyed the territory as if she were queen. She had been gentle and retiring, but from that day on she was the boss, the lead pony in the paddock."

Then came Ardmore Flyaway. If ponies have reputations for mischief, blame him and others like him. One day Dick looked outside and saw that the stallion was not alone in his paddock, where he should have been, but on his way to the next field with the mares. Dick put the pony back in his paddock, then went inside the house and accused Adele of forgetting to close the gate. Certainly not, replied Adele, who looked outside to see Flyaway once again about to enjoy mare company. Now it was her turn to accuse Dick. But their mutual indignation turned to wonder as they both watched: Flyaway deftly applied his hind end to the fence, and each time he did so the slide on the gate moved. Now and then he would interrupt proceedings to inspect the bolt. It took five or six minutes, but he got the gate open.

Eventually, the Rockwells had to deploy both vertical *and* horizontal controls on Flyaway's stall after he let himself out (using a similar tactic as before) to engage another stallion named George in battle. During the skirmish Flyaway had cut George's lip and by the time the two were separated, Flyaway — a pure-white Welsh pony — was covered in George's blood.

When not making trouble, Flyaway was winning ribbons all over North America. Flyaway is the only pony to have been named grand champion four times at the Royal Winter Fair. For five years he was the model North American champion at the National Welsh Pony Show in Devon, Pennsylvania. He sired half the champions whose ribbons adorn the pink room. Flyaway finally died in the spring of 1997, at the age of thirty-seven.

Thus do old ponies sometimes earn gracious retirements.

In October 1989, I was in southwestern Germany, among a small group of Canadians who were guests of a prominent German horseman. He and his wife lived rather splendidly in a country estate outside Stuttgart.

The hospitality, I remember, was both warm and formal. The elegant six-course dinner was served by a fleet of young men from the Lufthansa catering service, as tall and sleek as dressage horses. They kept our glasses filled with champagne or distinctive Swabian wines made from grapes grown on hillsides in Stuttgart itself, the bottles wrapped neatly in stiff white linen.

As the meal ended, our host exited the dinner table. Some minutes later we were called to the foyer, and there, wide-eyed and magnificent, stood the newest addition to the stable — a two-year-old colt pawing the Turkish rug at his feet and showing signs of agitation. I have no idea why the horseman felt compelled to bring the horse *inside* the house. Eccentric pride, I suppose.

We later toured the barn, where more fine horses were kept. Some were led out and run by grooms up and down stone walkways that we might see their fine action and conformation. But in one stall stood something immovable and gray, bearded, shaggy and ancient. He looked like a pony from a fable, a pack pony for elves and dwarves.

Here stood the childhood pony of our hostess, a fine dressage rider. She had felt too much affection for the pony to sell him when she had grown too big to ride him (likely three decades beforehand) or to put him down when he was beyond being ridden by anyone.

She had not forgotten the pony, once a nimble fellow and almost too much to handle. Nor should we.

Old western films often ended with a resonant bugle call — music to the ears of beset pioneers, their wagons drawn in a circle against Indians. The blue troopers gave chase to the painted ponies, closed in, and the warriors bit the dust.

The historical facts are otherwise.

Wagons never circled. And if the warriors chose to make a run for it, they almost always escaped; it was the troopers' horses who ate the dust. The scrubby little Indian ponies were exceedingly fast; the troopers' mounts were large, comparatively slow and heavily burdened.

Cavalry mounts typically stood sixteen hands high and weighed up to twelve hundred pounds. Half Thoroughbred, one-quarter draft horse and one-quarter mixed saddle stock, they were bred to carry a well-armed, self-sufficient soldier on long treks. Each horse had to bear the heavy saddle and the soldier in it, plus his weapons and ammunition, mess kit, canteen and rations, overcoat, blanket and bedsheet, along with the horse's halter, picket rope, feed bag and oats.

The typical Indian pony, on the other hand, averaged a little under fourteen hands and weighed about seven hundred pounds. The pony usually possessed a large head and little feet, heavy shoulders and hips and fine, strong limbs. The conformation, to pioneer eyes, was not the best, but the pony was sturdy and capable of great feats of endurance. His rider traveled light, with only a bridle and saddle pad, weapons, a small robe and a little dried meat.

And so it was that the "hammer-headed" painted pony, called a "cayuse" or "squaw horse" by scornful pioneers, almost always outdistanced the cavalry. Colonel de Trobriand, a cavalry officer on the plains, was clearly impressed. "The Indian pony without stopping," he wrote in 1867, "can cover a distance of from sixty to eighty miles between sunrise and sunset, while most of our horses are tired out at the end of thirty or forty miles." He found that "the movement of Indian horsemen is lighter, swifter and longer range than that of our cavalry, which means that they always get away from us."

By the end of the nineteenth century, warriors became farmers and ponies took a turn at the plow. Too light for the task, painted pony mares were bred to Morgans and Percherons, and soon enough tribal horses looked like cavalry remounts.

But in an even greater irony, the ponies who could deliver Indians from the hands of the white man could also, it turned out, deliver the white man's mail.

On April 3, 1860, while an animated crowd looked on, a bay mare left Pony Express headquarters in St. Joseph, Missouri, to begin the first mad dash across the continent. History does not record the pony's name, only that she was "bright," but she must have been eager to get under way. Hawkers and souvenir hunters, aware of what a historic — and potentially profitable — occasion this was, had been plucking hair from her tail to make rings and watch chains they hoped later to sell.

The freight company founders who launched the express service had seen a clear need. In the absence of a cross-country stagecoach line, it fell to pioneers to carry mail from one coast to another. When President William Henry Harrison died in 1841, four months elapsed before Californians got the news.

Pony Express mounts — usually, they were mustangs rounded up on the plains or ponies purchased from tribal owners — were small (fourteen hands or less) and often weighed eight hundred to a thousand pounds. The riders, short and wiry, about the size of jockeys, were astonishingly young — under eighteen years of age and some only fifteen. The company (headed up by stalwart Christians) made them swear never to drink to excess, curse or gamble, never to treat the animals cruelly and never to act other than as gentlemen. Anyone who broke the oath risked being fired without pay.

These young men had names that belonged in novels of the Old West: Sawed Off Cumbo, Deadwood Dick, Yank Little, and sometimes they went by only one name — such as Whipsaw or Boston.

One rider's name was Johnny Frye, and if you believe the tale that follows, then thanks to Johnny we have doughnuts. Frye rode a stretch between St. Joseph, Missouri, and Seneca, Kansas, and had made something of a name for himself as a jockey at local races. Several young women had noticed him and would wait for him along the way with cookies and cakes that he would grab as he raced by. But his admirers saw how he struggled to hold the desserts in one hand and to hand-ride (pump the reins) with the other. Why not, one young lady reasoned, put a hole in the middle of her baked offering? That

9.4 Pony Express: not the first courier service in history but easily the most celebrated.

way Johnny could slide a thumb through it, lean forward for a bite and still keep both hands on the reins.

Whatever Pony Express couriers ate, they took it on the fly. Riders changed horses every ten or twenty miles, carried no more than twenty pounds of mail and sometimes had to gallop 75 to 125 miles a day. The aim was to get the mail pouch — or *mochila,* as the Spanish called it — from St. Joseph to Sacramento, California — more than two thousand miles away — in eleven days. At some 153 stations along the way, the pouch would be hurriedly transferred from one exhausted horse and rider to a fresh team, and away. The names of some stations, such as Fort Laramie and Carson City, still sound familiar. Riders earned up to $125 a month and the cost of mailing a single letter was $5 — a lot of money then.

Despite the oath never to mistreat ponies, riders often rode for their lives and the ponies paid the price. There were the usual obstacles — deserts to cross, dust storms, rainstorms and blizzards, snowy mountain passes, clouds of biting insects, rattlesnakes, grizzly bears and cougars and, of course, war parties with designs on the Express pony and/or the rider's scalp. One rider, a young man named Joseph Wintle, was chased for many miles before he finally reached the relay station, where his pursuers gave up. The poor pony who had saved his life, however, died on the spot.

The danger along the road was often great, but so was the savvy of the riders. J. G. Kelly once took the *mochila* from another rider who had been shot by Indians and died shortly after reaching the station. But at Quaking Asp Grove, where his colleague had caught a bullet, Kelly put his trust in the horse: he took his rifle in both hands, dropped his reins on the horse's neck and sent him at a gallop through the pass. The trail was crooked, two miles long and only wide enough for a horse to pass through — a perfect place for an ambush. Kelly stopped only once — at the top of a hill to let his horse rest — and when he spied movement in the bushes he kept firing until it stopped, then he sped on. He made it, but soldiers who came through that same pass a few days later did not.

A lame or indifferent mount might have cost a rider his life, or many other lives, especially when towns under siege needed Pony Express riders both to deliver the mail *and* to fetch the cavalry. In Nevada in 1860, the Pah Utes around Pyramid Lake were blamed in a murder; half the posse sent to punish them were killed in an ambush. The fearful citizens of Virginia City converged in a stone house, and waited for the attack they believed was imminent. Those in nearby Carson City converted a hotel into a fort and likewise prepared for war. Help would only come if the Pony Express could get through.

When "Pony Bob" Haslam arrived in Carson City, he found every available Express pony out with the doomed posse. He fed his poor mount and coaxed another seventy-five miles out of him, but the rider at the next relay station was too frightened of the Pah Utes and refused to carry the mail. Pony Bob galloped on, and by the time he reached Cold

Spring he had gone 190 miles without a rest. The station at Cold Spring had been attacked, its keeper killed and all the horses taken. After a rest of only eight hours, pony and rider sped on. By the end of his marathon ride, Pony Bob had gone 380 miles and had been in the saddle almost continuously for thirty-six hours. Some may find it odd that we remember the name of Pony Bob but none of the extraordinary ponies who rode so long and so hard.

The record for long-distance riding among Pony Express riders, though, was claimed by William Frederick Cody, known later as Buffalo Bill. Young William started riding for the Express when he was fifteen. The son of a stagecoach driver in Iowa, he was by that time already a fine rider. Impatient with the short, safe runs he was assigned in the beginning, he yearned for the longer, riskier rides. He got his wish: the sometimes deadly 116-mile Wyoming run from Red Buttes to Three Crossings.

On one occasion, Cody arrived to find his replacement dead, so he rode another pony west and then other ponies all the way back east again — a virtual nonstop dash of 384 miles and just a shade longer than Pony Bob's epic ride.

By the time he was hired by the Kansas Pacific Railway to supply meat to its construction crews (he killed five thousand buffalo in eighteen months), Cody had made a name for himself. The legend of Buffalo Bill was born. His horse (called Lucrezia Borgia after a member of the Renaissance family whose name became synonymous with political ruthlessness) shared at least some of that fame.

Buffalo Bill soon discovered he could make a handsome living with his Wild West shows — entertainments put on in arenas for eastern folk wanting a taste of the Old West. Bill Cody simply played himself, or perhaps a grander version of himself.

From the 1880s until the early 1930s, these touring cir- cuslike shows featured mock battles between white actors dressed as U.S. Cavalry and Indian actors in full headdress;

9.5   Buffalo Bill: his wild west shows ensured his own fame and helped
      perpetuate the cowboy legend.

displays of horsemanship; and, of course, wagon trains under
attack. But because the arenas were round and action was
what the crowds wanted, the wagons moved in a circle. Film
directors saw this and almost every bad western since has had

wagons circling. As far as I can determine, no wagon master ever uttered the famous words "Circle the wagons!"

The Pony Express would only last a year and a half. Fast ponies were no match for the telegraph line in place by October 1861. A postal system using ponies had been a good idea, though hardly new. In ancient Persia twenty-four hundred years ago lived Cyrus the Great — the so-called King of the World, so vast was his empire. He may have been the first to establish a system of conveying royal messages by setting up posting stations fifteen miles apart.

A Greek traveler named Herodotus who saw the system operate was much impressed. "Neither snow nor rain nor heat nor gloom of night stays these couriers from swift completion of their appointed rounds," he wrote. His words would be adopted as the unofficial motto of the American post office, and are still etched in stone over the pillars of its New York City headquarters.

Genghis Khan, the fierce Mongol ruler, set up something similar — a courier system employing twenty thousand horses and ten thousand relay stations to keep in touch with the far-flung reaches of his empire. The ponies he used were grass fed and small, rarely more than thirteen or fourteen hands high.

Even a three-legged pony, it seems, has the wings of Pegasus. At the end of my day at Ardmore Stud, I bade farewell to Dick and Adele Rockwell. Standing in an entryway full of pony tack, I looked back at the Rockwells, the two of them framed by the door leading into the pink room. They might have been standing on a little rise, with one of those red-sky-at-night-sailor's-delight sunsets behind them. They had just told the story of Ariel, and it seemed a fitting end.

In 1960, a foal — the first to a pony mare called Ardmore Airborne — broke her right front leg above the knee when she was just three weeks old. Dick insisted to the veterinarian that some attempt be made to save her. At the time, it seemed a bold and reckless move.

In the clinic, Dick acted as the vet's surgical assistant as they anesthetized the foal, pulled the leg bone apart, inserted a pencil-sized pin into the marrow and fitted the foal with a walking cast actually designed for a Great Dane. Dick then had to change the dressing twice a day and lift up the foal to nurse at the mare. In time, the foal learned to lurch to her feet but needed Dick's help to settle back to the ground for rest. Eventually, the little pony devised her own method: she leaned against a wall and slid down. The smart pony by then had a name — Ardmore Ariel.

But the pin later began to protrude, forcing the vet to remove it, bandage the leg and confine her to her stall for three months. The leg muscle atrophied, and when the foal finally got the run of the pasture, she was not a pretty sight. As Adele looked on, the pony fell trying to follow her mother down a steep hill. Fell repeatedly and pathetically.

"We've done the wrong thing," Adele said.

The vet, however, held out hope that at least some leg muscle would return. Ariel did indeed learn to walk, trot, even canter in an odd sort of way. When Ariel was two, the Rockwells turned her out. After seeing her mother and other mares in a nearby pasture, she galloped toward a high fence, tucked up that wonky leg so it lay across — not under — her body and jumped the fence. Her mother greeted her effusively and licked her all over, especially that troublesome limb, as if to ask, "How's the leg?"

In her lifetime, Ariel would give birth to sixteen foals, twelve of them champions. She was finally put down in 1983 at the age of twenty-three when the hoof on the left front leg, the good one, began to cave in from years of compensating for the bad one.

In the years before she died, Ariel's older sister, Ardmore Skyborne, would chase the other ponies away to let Ariel drink in peace. "I've never seen a horse show such compassion," said Adele. But then, Ariel and Skyborne were not horses. They were ponies.

CHAPTER 10

HORSE TALES
TALL AND TRUE

The goat is a devil, the sheep is an angel,
the camel is a pilgrim, the horse is a hero.
TURKISH PROVERB

THE HORSE ON his way to the slaughterhouse is on a
journey of no return. But not always.

On November 29, 1988, a transport truck towing a pup
trailer on a Toronto freeway accidentally discharged part of
its cargo — a bay Standardbred gelding en route, with thirty-
seven other horses, to a cannery in Quebec. The horse — later
named Lucky — landed on his feet in more ways than one. He
skidded along the pavement, stick-handled across three lanes
of traffic at the height of rush hour, then leaped a barrier and
crossed nine more lanes. All astonishing, because Highway
401 at Yonge Street ranks among the least horse-friendly strips
on the planet.

After numerous close calls, the by then dazed horse had sustained only cuts and bruises. Two horse-wise drivers came to Lucky's aid: one man calmed him while a policewoman fashioned a rope halter, and they led him to safety. Lucky received veterinary treatment, and, better, a new lease on life when the Humane Society convinced his owner to hand him over. Eventually, the horse was adopted.

Thus did Lucky cheat the knacker. But the horse who dodges death or ignominy to become a champion — this is altogether too Disney-esque, too Walter Farley-ish, for words. Or is it? Let this chapter full of tall tales and near lies about horses begin with an absolutely true tale. Let it begin with the impossible story of Snow Man.

In February 1956, Harry de Leyer, owner of a small stable on Long Island and riding master at the Knox School for Girls, drove to the weekly horse auction in New Holland, Pennsylvania. On that bitter day, the skies shrouded in cloud, de Leyer sought a steady Thoroughbred school horse, but he had arrived too late. Only the dregs remained: fifteen sorry horses, bought for $60 each by a pet food manufacturer, were being loaded into a butcher's van. For someone like the Dutch-born de Leyer, who grew up with horses and loved them, it was a sad parade.

But from his position at the bottom of the ramp leading into the van, one horse caught his eye. A dappled gray. There was something about the lively tilt of the ears, something in his step — "a kindness in his sad eyes" is how de Leyer put it to me. The driver was packing them in tight, head to tail, and into that darkness de Leyer shouted.

"Hey, I'd like to see that horse."

"You crazy?" replied the driver. "He's just a horse, just another farm horse."

But the driver did as instructed and led the big angular gray back down the ramp and into the light, where the horse was revealed in all his misery. The burrs, the mud caked on the long heavy tail. The bite marks, sores and manure spots.

The missing shoe, the crease across his big chest from years of wearing a heavy collar and pulling a plow.

As he was being looked over he opened his eyes briefly, then closed them again. He was led back into the truck, but not for long; the driver would drop him off at de Leyer's farm. For $70, de Leyer had himself another horse, an equine mutt who looked to be half-Percheron and turned out to be nine years old.

Coming down the ramp at the farm, as de Leyer, his wife and five children looked on, the horse tripped over his own feet and stood there at the bottom in six inches of snow, blinking under a blinding sun. Snow Man, they would call him.

Five washings later, the accumulated layers of farm filth finally left his coat. The family groomed him, shoed him, put some weight on him, schooled him as best they could — and sold him that summer for $140 to a local chiropractor. But like Dr. Seuss's cat, he came back the very next day, at a gallop, and did so repeatedly no matter how high the enclosure. Two aspects of the horse became clear: Snow Man was loyal to de Leyer and Snow Man could jump.

De Leyer began to train him, at first with woeful results. The gray tripped over the cavaletti — poles on the ground that teach horses to lengthen and shorten stride — but he soon mastered the art and eventually it took only a cluck from de Leyer and even a riderless Snow Man would leap any fence. Like Big Ben, who took great joy as a colt in leaping fences when turned out in an arena, he was that rare horse who naturally, gleefully, takes to jumping.

In 1958, after humbling lesser competition, Snow Man jumped victoriously at Madison Square Garden in New York City and was later named American national champion two years running. Snow Man would win the puissance one minute (calling on his power) and the speed class the next (relying more on his quick turns than on pure speed). Offers to buy him ran as high as $400,000. But de Leyer had no interest in selling the Cinderella horse, as he came to be known. Countless

10.1   Snow Man swimming Long Island Sound with de Leyer children: from abbatoir truck to show jumping champion.

magazine articles were written about him, along with two books.

Back on the farm, he was a docile pet. Photographs in the November 7, 1959, issue of *Life* magazine show Snow Man being ridden by all five de Leyer children; another photograph has three of the kids on his back as he swims — apparently he *loved* to swim — Long Island Sound. His mouth is open, teeth showing. You would swear he was smiling. De Leyer, by the way, now seventy and living in Virginia, still rides and competes. "Horses do more than keep me young," he says. "They keep me happy."

Snow Man was not the only horse to leap from servitude to show ring. Laurens van der Post wrote a book earlier this decade called *About Blady*. Mystical in places and often autobiographical, the book is very much about human kinship with horses.

Van der Post and several friends were traveling through France decades ago when a woman in the car spotted a plow horse in a field. Headstrong and confident, she demanded that the car be stopped, then she strode purposefully out to the farmer behind the plow, bargained with him and bought the horse on the spot — over her husband's objections. Blady went on to become a great show jumper, and in one memorable competition, Blady and rider beat the gifted Spanish rider Paco Goyoago and his fine mount.

From plow horse to show horse. It happens.

Such miraculous transitions have their own appeal and represent one more aspect of the fervor humans feel for horses. If the passion for horses were seen as a great tree, the wealth of horse mythology would be the deep and tangled root.

At the other end of the tree, the top end, amid the light and airy branches, lies the fruit. Strange exotic fruit, tempting and often delicious to behold, but can you swallow it? Can you trust the taste? Is it hollow inside, or as genuine as the horse plum that grows wild on the North American continent?

Horsiness invites the trading of horse lore, the way cat or dog lovers do. We who value horses swap horse stories and offer them as proof — of horse character, stubbornness, intelligence (or lack of it), courage, devotion or high spirit. The stories say to those who prize horses, See? I told you so. Or, You won't believe this but . . .

One of the most enduring clever horse stories concerns a sixteenth-century horse in England called Marocco. Shakespeare's play *Love's Labor's Lost* has Armado engaging in banter with his page, Moth, who says, "Now here's three studied, ere you'll thrice wink: and how easy it is to put years to the word three, and study three years in two words, the dancing horse will tell you." The dancing horse was Marocco, a medium-sized bay gelding with a docked tail. Ben Jonson,

John Donne, Thomas Middleton and Sir Walter Raleigh all mention him in their work.

In his *History of the World*, published in 1614, Raleigh writes of Marocco's trainer, Thomas Bankes, a Scottish wine merchant and juggler, that "if [he] had lived in older times he would have shamed all the enchanters of the world, for whosoever was most famous among them could never master or instruct any beast as he did his horse."

In 1600, in a celebrated performance, Marocco climbed the spiral staircase of St. Paul's Cathedral in London. He could dance a jig and walk on his hind legs (forward, backward and in circles). It seemed he could also count. Someone would throw dice and Marocco would tap out with one silver-shoed hoof the number that came up. Shown the glove of a person in the audience during a performance, Marocco could even pick out its owner.

Clearly, Thomas Bankes was passing sly signs to Marocco. Credit the horse's attentiveness, the trainer's showmanship. But perhaps they were *too* clever. Obsessed with magic and witchcraft, sixteenth-century authorities exacted severe penalties for practicing either.

Unable to imagine such a gifted horse and trainer, audiences concluded that Bankes was a sorcerer and Marocco his apprentice. The trainer landed in dungeons several times. According to one story, he once ducked prison by having his (obviously holy) horse kneel before a man wearing a crucifix in his hat; Marocco then rose and kissed the cross.

But in Rome, the pope denounced both man and horse as devil worshipers. Bankes either escaped or was pardoned, because he did live to a ripe old age. Of the dancing horse nothing was heard again.

A hundred or so years later, the public reacquired a taste for trick ponies. In 1770, as mentioned earlier, Philip Astley launched the first of its kind — an equestrian circus in London that led to others in Dublin and Paris. Illustrations of the day depict his amphitheaters as circular arenas with magnificent

10.2   Sky Rider, Atlantic City, New Jersey, 1960: stunt-rider and horse
jump from a 75-foot-high tower into water.

chandeliers and four tiers of seating. They looked like opera houses, with horses, not singers, at the center.

Richard Lawrence, a veterinary surgeon who witnessed the entertainments at Astley's Theatre, was impressed by what he saw and the means of achieving it. "A good-natured clever man," he wrote, "may teach a horse anything, and it is a very mistaken notion that those horses which perform so many dexterous tricks at Astley's, and places of that description, are brought to execute them by violent means. The fact is, they are taught by gentle means only, and by rewarding them at the moment they obey; hence they become accustomed by habit to combine the recollection of the reward with the performance of the trick, and it becomes a pleasure to them instead of labor."

Astley's horses could dance a minuet, lie down on command and play dead during mock battles, not stirring until asked. Most famous of all was Billy, The Little Learned Military Horse. Like Marocco, he could apparently count (taking his cue from the clicking of his master's fingernails, as perhaps Marocco had done). As a favor Astley once loaned Billy to a colleague who sold the horse to pay off a debt. Astley spotted him later on the street, pulling a cart. The master clicked his fingernails and so spirited was the response that Billy almost overturned the cart. Horse and trainer were quickly reunited.

Billy could spell Astley's name in the earth with his hoof, distinguish gold from silver, ladies from gentlemen, and pluck a handkerchief from his owner's pocket. Asked to choose between death and fighting on the side of the Spanish, Billy dropped down, as if dead. Even in old age, Billy would, on command, wash his feet in water, remove his own saddle or lift a boiling kettle from a fire.

A century later, and along came a cart horse in Germany known as Kluge Hans, or Clever Hans. His story is still told by scientists to debunk notions of animal intelligence and, especially, to counter the human tendency to read too much

10.3   Wilhelm von Osten with his prize pupil, Clever Hans: smart horse,
        great expectations.

into animal actions. Clever Hans, these professors insist, was
not so clever at all.

I am less sure. At play here are two disparate elements: a
truly smart horse and humans ill-inclined to bestow the word
clever on any other creatures.

Hans came to be owned by Herr Wilhelm von Osten, a
Prussian nobleman who was impressed by how Hans backed
a cart along a circular driveway. You need only watch a skilled
transport driver back from a busy street into a narrow alley-
way to appreciate how devilishly difficult it is when cab and
rig, or horse and cart, are semidetached. Von Osten insisted
that horses in general, and this horse in particular, were
smarter than commonly believed.

A wealthy eccentric with the time to indulge his whims,
von Osten thought he could teach Hans to think, as a teacher

would a pupil. The one photograph I have seen of teacher and pupil together shows a bearded man in an ankle-length lab coat standing at the horse's shoulder. To his right are two blackboards busy with numbers and letters in chalk. Looming over the blackboards, a foot or so from the horse's nose, is an abacus.

The horse learned to respond to questions — by tapping his right hoof a certain number of times in mathematics, by selecting the correct wooden alphabet block or by shaking his head or nodding. Correct answers won him a bite of carrot; wrong answers earned dire threats from his master. Telling time, distinguishing coins, identifying musical scores, absorbing geography and politics all appeared to be within his grasp. Word spread. Clever Hans became world famous.

Was Hans a *thinking* horse? Or another Marocco? The world wanted to know. So on September 6, 1904, a committee of thirteen apparently smart people — including a zoologist, a politician, a veterinarian and a psychologist — set out to determine the answer.

They were at first amazed, then stumped. Hans, they were quite certain, was not actually doing the math, but if the horse was reading signs from von Osten, what signs?

Finally, Oskar Pfungst, a student at the Berlin Psychological Institute, proved that the horse was reading the body language of the questioner. The horse was lost for an answer when von Osten asked his multiple-choice questions out of the horse's sight; in his sight was another matter. Clever Hans — perhaps by observing his master's eyes or eyebrows, body tilt or rate of breathing or slight unconscious flaring of the nostrils — caught the clue.

The curious thing is how the world of science responded to this finding. A note of triumph was struck: a dumb animal had been restored to its rightful place below humankind. Even von Osten, who never meant to cue the horse, felt betrayed, as if Hans had somehow let him down. He died soon afterward, a bitter and disillusioned man.

I am not the first to ask the obvious question. (Vicki Hearne has also written about Hans.) Why did Clever Hans not receive more credit for his ability to read signs so subtle that thirteen learned observers could not detect them?

A somewhat similar case occurred in Virginia in the 1920s when a horse named Lady Wonder was claimed to be capable of reading minds. Someone would whisper a secret to the horse's owner, who then stood well back and looked on as the horse used her nose to move wooden blocks and spell out the secret. Many people expressed amazement, including an academic who published a glowing article about the horse.

But where von Osten was a completely honest man, Lady Wonder's trainer was a scam artist. A professional magician called in to investigate caught the trainer tapping the air with the whip in his hand when the horse hovered over the right block. Lady Wonder, like Clever Hans, was simply an acutely alert horse. But once again, high expectations were dashed and the horse, of course, got no credit.

Less well-known is the fact that a disciple of von Osten's, an affluent manufacturer named Karl Krall, who lived near the German town of Elberfeld, continued to teach horses. He even took sad old Hans into his stable. If we believe the reports of Maurice Maeterlinck in his book *The Unknown Guest*, published in 1914, Krall taught horses impressive feats of spelling and mathematics.

Maeterlinck, alone one day with a Krall-trained Arab horse, dictated the name of the hotel he was staying at. The horse, using his hooves to stamp out the alphabetic code he had been taught, correctly (one letter was wrong) spelled out the name. Maeterlinck credits a combination of equine intelligence and telepathy to explain his findings. Modern doubting Thomases like me dismiss this as simply too tall to be true.

I asked Vicki Hearne, a notable horse trainer, to respond. She neither dismissed nor endorsed the Maeterlinck claims: "You hang around dogs and horses and you come to realize that things go on. Animals read people so minutely and when people are coherent, you might as well call it telepathy." She remembered seeing footage of Colonel Alois Podhajsky, the famous trainer of Lipizzaner stallions. "He was an emblem," said Hearne, "of what it is to ride. I could hardly breathe as I watched that film. There were closeups as he did pirouettes and jumps, and you could not see his hand or leg move."

You might as well call it telepathy. The more we learn about animals, and the communication possible between and with animals, the more we are inclined to chip away at the human arrogance that for so long has kept Us and Them in distant camps.

In the late 1940s, the Union Milk Company in Calgary still delivered milk by horse-drawn wagon. A new horse, likely a massive Percheron or Belgian, with feathered feet the size of pie plates, had been given a route near the dairy so he could learn the ropes. At one point the driver ducked into a house to make a delivery, and when he returned he clucked to the horse.

10.4 Milk wagon in Berlin, 1909: the cart-horse has been known to diplay astonishing horse sense.

The horse refused to budge. Three times the driver slapped the horse on the back with the reins, but the horse stood statue still. Finally, the driver got out to investigate. Clutching the legs of the horse was a three-year-old child. The horse had not only tolerated this but, apparently out of concern for the boy's safety, ignored commands to proceed. This is a story about the horse as guardian angel.

Then there are stories about the horse as creature of habit. During the same era, at the Palliser Hotel in Calgary, a doorman daily offered sugar cubes to the milk-wagon horse passing by. One day the doorman fell ill and someone else took his place. When the horse arrived he stopped, and waited. What? No sugar? Up the steps he went, looking for the sugar man. He got halfway there before finally abandoning the plan. A photograph of milk wagon on hotel steps made *The Calgary Herald* the following day.

Horses, we know, like routine. For years at the stable where I learned to ride, my partner in the ring was a willing but often cranky mare named Cassie. The routine went like this: I would groom her; my instructor, Maria, and I would tack her up; I would ride her for an hour, walk her for a bit; and then came carrots — my fee for services rendered.

One day I brought house guests to visit the horses. I took carrots along, and the youngsters among us fed one to every horse. Every horse, that is, save Cassie. I was scheduled to ride her in an hour and her carrot would come *after* the ride.

We were not long in the ring before she did something she had never done before — she tried to buck me off! Possible explanation: what Cassie had observed was that every horse had gotten a carrot — except her. And now she was supposed to work for the architect of this insult? Forget it. Buck him.

From ancient Greece comes another story about horses and routines. The Greeks loved to race horses, as we do, and made horse racing an Olympic event. A mare called Aura threw her jockey early in an Olympic race but galloped on, just as horses in steeplechase races — even when the jockey is eight hedges back in a heap — often continue round the course. Aura went further: she did precisely the required number of loops around the track, beat all the other horses, then stopped in front of the judge's stand. To receive her prize, of course.

Then there are the stories that flatter us. These are stories, or so we think, of horses displaying affection, even devotion, to their riders.

Lucy Rees, author of *The Horse's Mind*, has coined a phrase to describe the bond between rider and horse. She believes that when you gain the trust of a horse, you enter that horse's society — his equine society. You become what Rees calls "an honorary horse."

J. Frank Dobie in *The Mustangs* sketches an ex-slave named Bob Lemmons who spent all his life gathering range mustangs. He did so alone, never in a gallop, and always by winning their trust until the herd actually let him lead them. "I acted like I was a mustang," Lemmons, then eighty-four, told Dobie in 1931. "I made the mustangs think I was one of them. Maybe I was in them days." When convinced he had been accepted as a senior member of a particular equine tribe, Lemmons would lead the wild horses to water, let them see him smelling for danger and then sleep in their midst with his own horse on a picket (a means of tethering a horse with a rope and stake in the ground).

The following story is told by a woman — perhaps an honorary horse herself — who offers it as proof of the protectiveness of horses, or at least of one horse in particular. And it reminds me very much of the story of Colonel — another draft horse — told at the book's beginning. The woman was in a corral with two horses when she suffered a dizzy spell and collapsed. One horse panicked and circled in a frenzy. At that point the woman lost consciousness. When she awoke, she was flat on her back and staring up, not into blue sky, but into the massive belly of the other horse, a Clydesdale mare.

The mare had created for the woman the only safe place in the corral; the mare's body and eyes kept the other, nervous, horse at bay. When the woman finally felt strong enough to rise, the mare lowered her neck, which the woman used as a lifting aid.

Books of the Old West recount similar stories. A cowboy on a cattle drive supposed to be on night watch nods off in his bedroll and wakes up to find the earth trembling and the cattle stampeding toward him. But the cowboy's own horse has straddled him, and the cattle race by like the wind around an old oak.

The horse-human fellowship has seen some astonishing displays of affection, some of them seemingly mutual in the extreme. In a book called *Talking With Horses*, Henry Blake describes an emotional reunion with his mare Fearless after his long absence for military duty.

He calls out to the herd and she comes at him in a full gallop with her ears back and her mouth wide-open, and for a moment Blake feels as though he's standing on railway tracks with an express train bearing down on him. But she skids to a halt about ten feet from him and proceeds to make every sign that she is thrilled to see him: "Then she took two steps forward and licked me all over from head to foot, and when she had done this for about three minutes the tears were running down my cheeks, so she thought that was enough of a good thing. Just to show me the status was still quo, she caught hold of me with her teeth and lifted me from the ground and shook me slowly backwards and forwards four or five times, then she put me down and rubbed me with her nose."

J. Frank Dobie tells the story of Prince, a strawberry roan sired by a French Canadian stallion out of a mustang mare. That Kansas horse saved his owner's life more times than TV Lassie ever saved Jeff, even led him out of the memorable blizzard of 1887. In his later years, the horse would bring the cows in from the pasture at day's end. Chester Evans, who owned him all this time, would hitch him to a wagon and send him to town, where someone would load the wagon with supplies, and then Prince would pull the supply-laden

10.5   The horse-human bond: what poet Maxine Kumin calls "the right
magnificent obsession."

wagon home. Prince died at the age of thirty-eight after a
freak accident. "No man," said Evans, choked with emotion
as he recalled his horse, "could have had a more congenial
companion or a truer friend."

It is a flattering thing to be loved by an animal. I can let myself be convinced that our dog loves me, too. The face licking,
the tail wagging, the way she looks up at me when we're driving in the truck and her chin is resting on my right thigh. Is
that not pure adoration? Yes, but the minute another dog
enters the picture, I am second fiddle, or second Fido. As one
expert has opined, the prime interest of dogs is other dogs.

Certain horse books contain similar reminders to check
our more romantic notions. In *The Nature of Horses*, Stephen
Budiansky observes that his own horse behaves one way at

home (where only one pony offers him "equine company") and another way at a large riding stable with fifty other horse neighbors.

At home, says Budiansky, the horse is often quite affectionate and apparently eager for human contact in the paddock and at the fence line. At the stable, on the other hand, Budiansky and his children are more inclined to get the equine cool shoulder. "Our horses' affection for us, their owners," he says, "is unquestionably real, grounded in a basic instinct to form friendship bonds. It is slightly bruising to our egos, though, to realize that they bond with us only for lack of better company."

The horse-fevered do not dwell on such realities. We prefer tales like the one told to me by Anne Zander, who rides near Cookstown, Ontario. Zander, who had ridden as a youth, years later acquired a husband and a farm, and her zeal for riding returned. She only lacked a horse.

Enter Dr. Broom, a sixteen-three-hand Thoroughbred with impressive training times and bloodlines (by Bold Ruckus, a grandson of Bold Ruler). Despite that, he had been banned from the track for refusing to leave the starting gate. "All redhead with an attitude" was how Zander described him in a letter to me.

Professional trainers had washed their hands of the wired horse, so what luck would an amateur have? No luck at all, it seemed. Zander hauled that horse to shows, but only after spending hundreds of hours simply teaching him to enter a horse van. The horse kicked and bit, stepped on, lashed at, bucked off and ran away with his rider. In the fall of 1994, she took him to an entry-level horse show and the good doctor — Broom, that is — kicked Anne's husband, Claus, so badly he had to be treated by ambulance attendants and later wound up in hospital. Anne rode Dr. Broom in the trial, but they were eliminated in both the stadium and cross-country events. Though the jumps were small enough to step over, Dr. Broom would not deign to go near them.

As determined as she was apparently daft, Zander stuck it out with Dr. Broom. One day the prominent New Zealand rider and trainer Mark Todd — a two-time Olympic gold medalist in eventing — was giving clinics in the area, and Dr. Broom made his acquaintance. In the saddle, Todd took the measure of the horse and pronounced him one who "likes to get a bit uptight." Zander laughed at the understatement, but those few minutes of schooling by a skilled and sensitive rider may have launched a turnaround.

In 1996, Dr. Broom consistently went clean in both cross-country and stadium rounds. "He was no longer," wrote Zander, "a $5 horse." The real test was yet to come. While haying that year, Zander injured her back and required surgery. She lay abed for weeks and imagined Dr. Broom in his stall — an alarm clock wound tighter and tighter and set to go off.

When Zander was finally released from hospital, her back still ached and her left leg felt useless, but she was bent on riding. And how did Dr. Broom react to his essentially disabled rider?

"My wild horse," says Zander, "was a perfect gentleman. He stood stock-still for me to haul myself up. He did not move until I asked. I had no balance and had to hang on to his mane. If I started to lose my balance, or the pain was intolerable, I just said 'Whoa' and he would stop. We tried a few trot steps and a few at canter. He was a letter-perfect Pony Club horse."

Later, her coach mounted "Doc" who promptly tried to buck him off. Anne is left amazed by all this, and grateful. "He can still be a red-hot chestnut, but he came through for me at a time when I needed him most."

Though mechanized horsepower has supplanted the horse-drawn wagon in many places globally, a few eccentrics in the Western world, such as Don and Vi Godwin of Mulvane, Kansas, still put their faith in hay-burners.

On October 3, 1982, the retired couple took to the road

in a pickup camper turned backward and set on a four-wheel farm trailer, the whole thing pulled by two Belgian draft horses, full brothers named Barney and, well, Brother. By 1993, when they finally ceased wandering, the Godwins had traveled more than eleven thousand miles crisscrossing the continent. They typically covered twenty-four miles a day (top speed for "the boys," as they called the young chestnut horses, was four miles an hour). Daily fuel consumption amounted to one bale of hay and a bucket of grain.

Don Godwin had worked delivering mail and as an auctioneer, but all his life he had trained draft horses and nurtured a childhood dream of crossing the country in a covered wagon behind horses "with flaxen mane and tail." His great-uncle, a horse trader who toured the Midwest in a horse-drawn wagon, once took six-year-old Don with him for a day and a night. That planted the seed. Retaining a piece of land in Kansas as a home base, the Godwins sold everything to finance the voyage. Because they traveled at such a leisurely pace and because the sight of a horse-drawn trailer invited conversation with strangers, they met new people every day.

Now sixty-seven, Godwin told me the horses were virtual passports on the road. "There are a lot of horse lovers out there," he said. "Those horses got us into more homes and caused us to meet more people . . ." Locals driving by would see the horses, brake, and pretty soon the brother Belgians had hay and grain laid on, the Godwins had a meal and local radio stations had their lead news item — thus paving the way for more generosity down the road.

Godwin finally sold his beloved Belgians to a fellow in Nebraska, where he visits them on occasion and delights when they recognize him and come to his whistle. The trailer went to a man in Alberta with a similar notion of horse-powered travel. It seems the man had written a letter to *Draft Horse Journal* seeking practical information. A magazine staffer put him on to Godwin, and the deal was struck.

"You've got to love horses," Godwin says to anyone pondering a similar pilgrimage. "And you've got to have the right two people." The Godwins come from large families (thirteen in hers, nine in his) and were taught resourcefulness during the Great Depression, along with the saving grace of wit, which he clearly still possesses. "When we were kids," he explains, "we ate so much wild rabbit that every time a dog barked we ran under the porch."

Set aside your fear, Don Godwin advises, and have a little faith in your own vision. He and Vi have written a book about their travels, with proceeds going to their church. The book's title? *Faith vs. Fear.*

Fearless, too, are the Grant family from Scotland, who left home in 1990 to circumnavigate the globe in a horse-drawn caravan. David Grant, his wife, Kate, and their three children, Torcuil, Eilidh and Fionn, now eighteen, sixteen and thirteen, respectively, crossed Europe, Asia and North America before flying home from Halifax in the fall of 1997.

When I talked to him early in 1998, David Grant was sorting out the fourteen journals that will form the basis of what promises to be an extraordinary book. A central character in that book will be Traceur, the seventeen-hand, eighteen-hundred-pound Breton-Percheron cross who took them most of the way around the globe (two other horses did stints at the beginning and end). Grant described him as a gentle, temperamental bay gelding who in six years was never once tied up at night and was always there in the morning — save one time in Mongolia when Grant "got cross with him" and Traceur, perhaps sulking, repaid him by wandering off that night.

"He died of a brain tumor in America," Grant recalled, "at the Missouri River. We had gone ten thousand miles with him, and when Eilidh found him in the morning it was horrendous for her. She and I were closest to that horse. It was like losing a member of the family. The only time on the whole trip I considered quitting was when Traceur died."

A breezy man with a quick wit, Grant said the seven-year odyssey was occasioned by "cold, wet Scottish winters" and seeing, in a tourist magazine, a picture of a skewbald horse pulling a round-top Gypsy wagon through the Irish country-side. They took just such a trip in Ireland as a test run and embarked soon afterward.

The Grants arrived home, many years later, to no home, having sold theirs to finance the trip. But they had stories to tell — of the Balkan War, of drunken Mongolian horse thieves, deportation from China and encounters with a splendid array of people.

Grant's children learned, firsthand, lessons in geography, politics and history from living the gypsy life. "I'd recommend it," Grant said. "Sure, do it. Do it while you may."

Horse sense is a mysterious thing. Years ago a cart horse in Spain stubbornly refused to enter a mountain tunnel she had passed through countless times before. Traffic backed up behind her, and you can imagine the driver's anger and the stern tactics employed to move her. When the tunnel collapsed shortly afterward, the mare was hailed as a heroine.

Why did horses in San Francisco on April 18, 1906, thrash around and break free of their stalls hours before the earthquake struck? How to explain the many documented stories of horses getting their riders safely home through unfamiliar jungles and in winter whiteouts?

A horse being ridden at a gallop across a thin cover of snow may suddenly leap, as if over an invisible brook. Returning to that spot to know why, the rider discovers a groundhog hole. How did the horse detect it, and, stranger still, in a full gallop? Was it the sense of smell, or some other faculty beyond our ken?

Let's revisit that incident in *Black Beauty* when the horse balks at crossing the bridge. Beauty himself narrates the book, of course, and though there is no real explanation offered, he, like the Spanish cart horse, is determined not to go forward. Beauty had an inkling: "The moment my feet touched the first part of the bridge, I felt sure there was something wrong."

Was it a sixth sense, or something more actual? The editors of *Equus* magazine in the United States have published numerous reference guides, including one called *Understanding Equine Behavior*. In that little book, the authors consider the scene of Beauty at the bridge and list any number of equine sensitivities that might have offered fair warning. The storm itself would have put the horse's entire nervous system on full alert. The bridge, though familiar, would still have commanded the horse's respect because of its vibrations and hollow footing. Perhaps the break in the middle of the bridge altered the usual give and sway of the timbers, which might have been creaking. Black Beauty's night vision, abetted by his memory of previous bridge crossings, might have spotted a change in the angle of the rails. We can only guess at what might have informed the horse, but such radar is likely both real and a great deal more complex than humans have ever imagined.

Lacking a clear explanation, we put it down to horse sense. That said, we might also be on the lookout for horse manure.

Did some streetcar horses earlier this century *really* stop outside the offices in New York of the ASPCA (American Society for the Prevention of Cruelty to Animals) until an officer came out to alleviate their suffering?

Did Dr. Le Gear — claimed to be the tallest horse in the world — *really* exist? The Percheron gelding, who died in 1919, was listed at twenty-one hands and was said to weigh 2,996 pounds.

Did a horse called Ben Holt *really* jump nine feet six inches at the Sydney Royal Horse Show in 1938?

Was Topolino, a Libyan-born mount who served in the Italian army, *really* fifty-one years old when he died in 1960?

And during the Mexican Revolution earlier this century, could a palomino called Canelo ridden by a Mexican general *really* smell Pancho Villa's rebel soldiers fifteen miles away?

The word *canelo* means, in Spanish, cinnamon — too dark for a palomino and perhaps our first clue that the story is not to be trusted. Allowed to graze unhobbled around the camp, Canelo would sense the enemy and then, good bodyguard that he was, awaken the general (Esteban Falcón) by nipping him and pawing the ground.

One night, the story goes, the general was inside a house. It must have been a very large "house," for it also contained Canelo and a three-hundred-man army. At 2 A.M. the horse began to whinny, and the general, thinking him hungry, had a servant bring him corn. But the horse ignored it and finally kicked open the door. The general, perhaps anticipating *Lassie* movies of the future, somehow understood Canelo's message; he and his men saddled up and waited on a nearby hilltop. Some six hundred rebels were killed in the ensuing ambush.

Equally strange is the story of Fred Kimball, a California horse psychic used by many horsemen and -women, including some of the top-ranked horse trainers and riders in North America. Anne Kursinski, who rides for the U.S. equestrian team in show jumping; the former Lisa Carlsen (now Lubitz), who has ridden for the Canadian team; and Barb Mitchell, an outstanding Canadian trainer, have all used Fred Kimball.

Born in Massachusetts, Kimball worked as a sailor most of his life. He traveled the world and met monks and sages in Tibet and India, where he learned all he could about mind reading, an art he claimed to have mastered. Until he died in 1996 at the age of ninety-one, riders would call him up at his home near Idyllwild, California, give the name and sex of their horses, and Kimball would diagnose the horses' physical or psychological ailments. The fee was $25; he trusted clients to pop a check in the mail.

Riders remember him as odd but often accurate. A few years ago, a horse bucked off a rider at an eventing competition in Orillia, Ontario, and galloped away. After three days, all attempts at finding the horse had failed. Call "Indian Fred," someone said. (Though he had no Indian blood in his veins, that appellation persisted.) Kimball described a particular valley by water and said the horse was suffering with a shoulder problem. A local person identified the place, and there was the missing horse, all right — sore shoulder and all.

Nancy Page, a rider in Battersea, Ontario, told me she called Fred Kimball on a dare. Though he was quite ill and would die soon afterward, he told Page many things about her horses that she found eerily correct and she felt her skepticism draining away. In essence, Kimball claimed to converse with each horse and then reported the burden of that conversation to the owner.

Page learned, for example, that her two-year-old colt Brady took great pride in his feet, for he fashioned himself a jumper one day. Right after her chat, Page went to the paddock and brought Brady — a horse she had raised as a foal and who was as loyal as a dog — into his stall. "Nice feet, Brady," she told the horse in a mocking way, whereupon the horse, who had been quietly eating his hay, lunged at Page and bit her on the shoulder. He had never done anything like that before and has not since. Page, to be safe, has since refrained from making inflammatory comments about Brady's feet.

Help for horses can come in all shapes and sizes. Some horses, especially racehorses, walk nervously in their stalls and need "mascots" to calm them and keep them company. Exterminator, the 1918 Kentucky Derby winner, grew attached to a Shetland pony named Peanuts. Over the course of twenty-one years, the lean and lanky gelding had three such ponies (Peanuts every one). He loved them all and mourned them when they died. A famous racehorse of the eighteenth century, a major player in Thoroughbred genealogy called the Darley Arabian, befriended a cat who used to sit on his back in the stable or nestle against him. When the horse died, the grief-stricken cat refused to eat, until he, too, died.

Native Dancer, the splendid "gray ghost" of the 1950s, also displayed a great fondness for cats, and in particular a black stray who wandered into his railway boxcar during one of his many road trips and eventually became his stablemate.

Examples abound of horses — social creatures, after all —

bonding with humans and even other animals. I read of an Appaloosa who adopted a certain chicken, shared meals with the bird and forbade other chickens to seek her company. Racehorses have bonded with goats, dogs, rabbits, potbellied pigs, burros and roosters. A Thoroughbred named Hodge, who ran in the 1914 Kentucky Derby, loved a talking crow who used to sit on a fence along the backstretch and yell, "Come on, Hodge! Come on, Hodge!"

In the world of horses, the true tale is only a horse tail removed from the tall tale. Just hanging around stables, I have heard stories about horses who could slip the bolts on their stalls, slip Houdini-style out of warm-up blankets; horses who loved to eat hats and buttons, or mischievous horses who would hide brushes and other grooming tools in the straw. I have heard of horses who removed ladders, leaving their owners stranded on a roof. A rider told me of a horse who bites on the halter in his rider's hands, hoping to engage her in a little tug of war. A Canadian Thoroughbred called Le Danseur used to grab a rub bag and play catch with his groom.

Nicolette Engelman of Strongsville, Ohio, tells the story of Deter (short for Determined Effort), a sickly foal whose grand sire was Determine (the 1954 Kentucky Derby winner). Given little chance of surviving, the small black colt somehow pulled through. His stablemate was a six-month-old Rottweiler called Rolf, whose company he much preferred over horses.

The horse would kick a stall ball with his hooves, and he and Rolf would chase it. They played tug of war with dog toys. More canine than equine, Deter did everything but bark.

If Deter was a horse who thought he was a dog, Whippoorwill Hello! was a horse who thought she was a human. Mary Jean Vasiloff of Old Lyme, Connecticut, tells of this foal born unexpectedly in February of 1968. Vasiloff found the almost frozen foal in the morning near a distressed and milkless mare. After a week in the vet's incubator, the sickly orphan took up lodgings in Vasiloff's house. Vasiloff fed her every fifteen minutes, day and night, for a month.

It took a mere three days to housebreak the little horse. She would proudly pull at someone's sleeve, then run to urinate in her stall, with its rubber mats and deep bed of shavings. They called her Whippoorwill Hello! in part because the farm was on Whippoorwill Road and in part because she whinnied every time the phone rang. The foal played inhouse tag and soccer with Vasiloff's sons, drank water from the bathroom sink and chased the cat and dog under the couch.

No surprise, the filly was difficult to train. Convinced her place was in the home and not in the pasture, she hated other horses. Eventually, she permitted some, but not all, children to ride her. "She allowed the slow-witted, the lame or the timid," said Vasiloff, but let any able-bodied, experienced rider on her back and "she was instantly the Horse from Hell."

When the then fledgling North American Riding for the Handicapped Association launched a local chapter, Whippoorwill Hello! was donated to the program. She helped dozens of disabled children learn to ride, qualified one rider for the special Olympics and even performed on the White House lawn with Jim Brady in the saddle. (He was the presidential press secretary shot in the head during an attempt on Ronald Reagan's life in 1981).

Eventually, Whippoorwill Hello!, finally content to pasture with other horses, was stabled at a farm in Virginia. "One day," says Vasiloff, "as she was cantering through the pasture with her friends, she ran into a pile of branches, puncturing her lungs and heart. She died from the injury, and on autopsy they discovered that Whippoorwill Hello! was blind, and had been all her life."

No one can deny the extraordinary feats of cutting horses, those marvels of western ranches. Quarter Horses, many of them, are neck reined — to turn left, for example, the rider simply applies the rein to the right side of the horse's neck. Cattle drives require horses capable of quick stops, tight turns and rapid acceleration. The Quarter Horse is brilliantly suited to the task. How brilliant? Cowboys used to engage in competitive storytelling.

One yarn has a riderless lone horse working a herd of fifteen hundred cattle. After he had cut out one big steer three times, only to see him rejoin the herd, the cutting horse wonder finally lost his temper: he grabbed that steer's tail in his teeth, gave it a twist and him a somersault. The horse then sat on him for ten minutes.

Finally, there was the tale of the cutting horse who was so good, his rider bragged, he "could cut the baking powder out of a biscuit without breaking the crust."

Other stories serve to remind us of the powerful and intricate bond that sometimes exists between horses themselves. The wild-horse herd, for example, is like an extended family, but not one free of vices. Stallions may rape mares and kill colts after a harem takeover. Yet cohesiveness and even altruism are common. The story is told of a mustang roundup in Saskatchewan earlier this century in which a chased mare ran alongside the stallion, her muzzle always at his flank. They seemed inseparable and only after capture did the truth come out: like Whippoorwill Hello! she was blind.

Doreen Freer tells the story of Jube the Wonder Horse, as they called him in western New York State. Like a stunt horse in a B western, the Arab stallion would sit down, lie down, roll over, nod and shake his head for yes and no, even take bows. Some days he would do all this in his pasture — simply to amuse onlookers.

Shortly before the chestnut horse died, Jube was pastured with his three-year-old son, a paint named Murphy. "You could see the two of them running side by side," Freer recalls, "tossing their heads, their tails way up and their necks arched with pride. It was a daily ritual, the two of them playing and later, a nap for Jude, always in the same spot." One day he dropped heavily and never rose again.

Murphy pawed at his father, as if imploring him to play, then took Jube's mane in his teeth and tried to make him rise. But the head fell back slackly. At that point every horse in the pasture gathered around Jube and stayed there a full hour — "as if to mourn the passing of an old friend." The last to leave was Murphy.

The ritual mourning seems not to have been an isolated case. D. Lovell Coombs, in the Ottawa Valley, tells the story of Royal Mandy and the manner of her dying. After three foals and a thousand rides, Coombs felt he owed her "the familiar touch of a friend and a goodbye without terror."

The needle was given, and when Royal Mandy dropped, the other horses fidgeted and her longtime stablemate began a panicky gallop around the pasture. The backhoe was brought in to bury her, and the despondent Coombs walked heavily to the house with the vet. When he next looked out to the pasture, he marveled at what he saw.

Mandy's three pasture mates had formed a semicircle around the freshly dug grave. They stood that way, heads bowed, utterly still for half an hour. "If I ever doubted animals have feelings and compassion and a place in God's world," wrote Coombs, "I lost those doubts the day I saw three horses stand silent vigil over a lost friend."

Mandy was buried in a pasture with few daisies. The following spring, hundreds of daisies — those wildflowers with golden hearts and white petals — bloomed where Mandy lay.

Sometimes a horse story, ostensibly about human-horse affection, bears a hint of something else. The horse possesses a special quality and the rider seeks a share in it; that desire to partake of the horse's greatness strikes me as an integral aspect of humankind's intense attraction to the horse. The story is told of Jamal, a Bedouin who owned an Arab mare with legendary speed. He spurned all offers to buy her, guarding her obsessively, and when the governor of Damascus offered a horse's nose bag full of gold for the man who stole her, Jamal grew even more wary.

Despite elaborate precautions, Gafar, member of a tribe hostile to Jamal's, did manage to steal her in the night, and before galloping off he woke Jamal and boasted of his feat. Leaping onto his brother's mare, a sister horse to his, Jamal chased the thief and was actually gaining on his own horse when he said to the thief, "Pinch her right ear and give her a touch of the left heel." This was the key and the mare shot off into the darkness.

"You father of a jackass," his tribesmen later mocked Jamal.

"I would rather lose her than sully her reputation," Jamal explained. "Would you have me suffer it being said among the tribes that any other mare proved fleeter than mine? It still is true. She never met her match."

EPILOGUE

*No one ever came to grief — except honorable grief —
through riding horses. No hour of life is lost that is spent
in the saddle. Young men have often been ruined through
owning horses, or through backing horses, but never
through riding them; unless of course they break their
necks, which, taken at a gallop, is a very good death to die.*
SIR WINSTON CHURCHILL

He who would venture nothing must not get on a horse.
SPANISH PROVERB

To the question, Why are some of us wild about horses? there is no one answer. But the more I inquire, the more I trust my instincts. Certain responses ring vividly true.

That of Vickie Rowlands, for example. Her story — pony rescues foal from murderous gelding, told in chapter 9 — derives from a life in the company of horses. She sleeps in their stalls at foaling time, mourns them deeply when they die, rejoices at their victories and, as the operator of a horse-trek business in rural South Africa, sees to it that each horse is bedded down at the end of the day.

"The most amazing thing about horses," she told me, "is that creatures so powerful, so capable of destruction — " and here she began to speak slowly and in italics " — *let us ride them*." That, finally and more than anything, is what sustains the horse-human bond: the generosity of horses induces in riders a pure, largely unconscious gratitude.

Bill Barich, who hung around the racetrack for months before writing *Laughing in the Hills*, reflects on that generosity as he ponders its clear opposite: the terrible injuries that horses occasionally inflict. "Every groom, and almost every trainer," he wrote, "told such tales, of hoofs flying out of nowhere to bunch an ear like cauliflower or scatter teeth like Chiclets . . . There was really nothing to protect you from the horses except a sort of grace conferred by the animals themselves." Grace is a useful notion here. So is faith.

Riding a horse or even getting close to horses — to pick their hooves, groom them, saddle them — is an act of faith. In Wyoming, I was leading Radish and another horse to water when I carelessly let the lead shank lengthen and the rope caressed Radish's flank: he exploded, swung round, kicked blindly. Pure reflex spared me harm, and despite the words of the smiling wrangler — "You handled it like a pro" — I was reminded of how a dude's head can split like a cantaloupe.

The cowboys' apparent casualness around horses, I now know, masks a constant vigilance. Let your guard down even for a second and you may pay a price. But even those hurt by horses, it struck me, rarely lose faith.

Christopher Reeve, an actor best remembered for his role in the *Superman* movies, fell hard from a horse in 1995 during an eventing competition in Virginia. Barring a medical advance, he will pass the rest of his days in a wheelchair. Every year in the United States, some fifty thousand people are injured riding horses, and some two hundred die from those injuries. But a *New York Times* reporter went around to the stable where Reeve fell and asked riders, Has the accident given you pause? The answer, invariably, was no.

One Virginian in his late sixties, paralyzed from the waist down after a fall from a horse and forced to use braces and a walker, was back in the saddle two years later. "Riding is therapy," he told the *Times* reporter. "I figure, the horse got me here, and it's going to get me out of here."

In the fall of 1997, I embarked on a pilgrimage around New England to meet three wise horsewomen, hoping to get a better grip on why humans are attracted to horses. I aimed to visit the horse farms of a professor, a poet and a philosopher. It just so happens that all three women are published poets and philosophers by inclination and teachers by trade. Each in her own way has built her life around horses and has written intelligently and widely about them.

First stop was the home of the professor. Golden Dream Farm, not really a farm but a small acreage named after a beloved horse, lies near Adamsville, Rhode Island. Set back from a quiet country road, the modest house sits atop a gentle hill where a homemade rubber horse hangs from a tree for grandchildren to ride in lieu of the obligatory tire. A silent little Shetland sheep dog welcomed me politely and discreetly at the door.

Elizabeth Atwood Lawrence is a woman ahead of her time. Born in 1929, she spent the first fifteen years of her working life as a practicing veterinarian, the latter seventeen years teaching cultural anthropology at the Tufts University School of Veterinary Medicine in Boston. And during those thirty-two years, Lawrence has witnessed a sea change in the way humans perceive animals.

"At vet school," she remembered, sitting on her couch with a cat on her lap and walls of books all around her, "if I had broached the subject of human-animal relationships, they would have thought me crazy. I always tell my students how lucky they are. We now teach vet students about dealing with the grief of clients who have lost a pet. That was unheard of years ago." Lawrence's own course at Tufts on human-animal relationships was the first of its kind and the first to be made part of the core curriculum, though other universities now offer similar ones.

"Animals," she told me, "are my life." But it is the particular relationship between horses and humans that has long preoccupied Lawrence. Her books — among them *Hoofbeats and Society: Studies of Human-Horse Interactions; Rodeo: An Anthropologist Looks at the Wild and the Tame;* and *His Very Silence Speaks: Comanche — The Horse Who Survived Custer's Last Stand* — all strive to decipher why the horse holds such a pivotal place in the human heart.

Lawrence laughed as she recalled the little war she had had with her academic publisher over the subtitle of the Comanche book. They had wanted it to read *The Horse That* and not, as Lawrence insisted, *The Horse Who.* For eons, language denied animals what she called "personhood." A horse was an "it" and not a "he" or a "she." Now even that is changing. Throughout my own book, I, like Lawrence, have conferred personhood on horses.

If we ride or own horses these days, said Lawrence (as her cat left her lap for my notebook), it is because we want to — may, in fact, *need* to. "Humans have let each other down," Lawrence argued. "Our society is so fragmented. I sometimes go to little towns and I see there the same wonderful sense of community that I see in preindustrial people — the ones who are left. If our needs were met by other humans, I wonder if we'd turn to animals the way we do. It's about kinship. Even when relationships with other humans are satisfying, we still seek out animals because they offer us something that humans can't."

When I asked Lawrence what the horse, in particular, offers in terms of kinship, she as much as said, as Elizabeth Barrett Browning did, "How do I love thee? Let me count the ways." When Lawrence thinks of the horse, she thinks of pleasant rides in New England woods, and a connection with nature. (The time her recently acquired Morgan mare, Easter

Bonnet, shied and did *not* run off and leave her grounded rider was the beginning of her bond with that horse.) The power of the horse, and the risk implied, has immense appeal: a horse can buck and tear across a paddock, then brake by the fence for a child's outstretched hand. The transformation can occur in seconds. How the horse embodies both the wild and the tame — this was much the focus of Lawrence's rodeo book.

"The fine-tuned communication between rider and horse," she once wrote, "is both physical and mental, as the beauty and grace of the horse's movement become qualities possessed by the rider. Even for people who do not ride, horses represent freedom, power and romantic beauty."

In our mechanized age, said Lawrence, we need the horse's spirit. The horse fills us with nostalgia for a simpler time. Her own field research told Lawrence that while a cop in a cruiser is just a cop, a cop on a horse is seen as a blue knight. "No one," she once wrote, "pats a cruiser." Riding, she believes, keeps you young — in a physical way and in a deeper way. Lawrence described a study of British geriatrics who bored their visitors with their ailments until the oldsters were given pet budgies, whereupon the birds became the focus, to everyone's delight. "Horses," said Lawrence, "are in that category, too. The horse takes you outside yourself." Finally, there is the horse's undeniable beauty. Lawrence uttered six words to express her admiration: "I just love looking at them."

Stop two. The poet's house. Pobiz Farm. Down at the end of Parade Ground Cemetery Road, near Warner, New Hampshire. The "poetry business," Maxine Kumin's name for her calling — writing and teaching poetry, reading at far-flung literary festivals — brings in enough income to let her breed and train horses, the first of them an abused horse called Taboo, rescued in 1974 from a trip to the abbatoir. Kumin is very good at both poetry *and* horses: the Pulitzer Prize she won in 1973 will attest to the former and I will attest to the latter.

We were walking up a steep hill on her 175-acre farm to get to the flat where she works her horses. The late-October sun was uncommonly kind; the leaves, still luminous. At seventy-two and bothered by arthritis, she no longer enters endurance riding events as she long did. "It grieves me," she said, "that I can't do it anymore." But she still rides, still competes in combined driving events (to the horror of her children, for this horse-and-buggy sport can be dangerous). I confessed to Kumin, sheepishly, that I do not actually *own* a horse. The look she gave me on the hill was all at once quizzical, maternal and sympathetic, and she stopped in her tracks to deliver it. "How do you *stand* it?" she then asked.

Maxine Kumin has written almost two dozen books of poetry, short stories, novels and essays. The horse, especially communicating with the horse, is a prominent theme in most of them. In *Looking for Luck*, she wonders, "Perhaps in the last great turn of the wheel / I was some sort of grazing animal . . ." *In Deep: Country Essays* contains the revelation that manure mixed with wood shavings smelled better to her as a child than perfume, and still does.

In *Women, Animals, and Vegetables*, she describes her childhood obsession with horses. She rode at a stable for $1 an hour or in exchange for stall mucking but never actually owned a horse until she was in her forties and bought Welsh ponies for one of her daughters. The derelict farm in the Mink Hills then looked nothing so neat and attuned to the seasonal round as Pobiz Farm does now.

There is a stillness about the place: we talked on a little patio facing the nearby barn and paddock, and the sound of dry leaves skittering in circles at Kumin's feet seemed singularly fresh, as if I were hearing it for the first time. At one point during our chat, a bird called from the banked forest that surrounds the farm and she immediately recognized the song as that of a cardinal.

My sense is that Kumin and her husband, Victor, know every inch of the farm they bought in 1963 and made their permanent home in 1976. At the heart of that profound sense of place is the horse.

"Horses have been our salvation," she said, kindly shoo-
ing two dogs who had come to me for a pat. "The horses keep
us young. We're both so physically active. Victor is seventy-
six, and he still rides his brood mare, who's twenty-one and is
still opinionated, single-minded, twitchy and wonderful. We
had guests from Texas the other day, and they observed that
our four horses are our children now. That's certainly true.
The bond is very, very tight. Maybe it's because we live in a
technical age, but the horse seems so tactile, so physical and
emotional, not programmed or mechanized. The horse is different every day, and a day without horses is a lost day."

Like Lawrence the professor, Kumin the poet finds in her
four horses (part Arab every one), a connection with nature.
Both lean and seemingly ageless, the Kumins are still carving
out trails in the rock-strewn forest. Maxine Kumin loves the
ritual, "the dailiness," she called it — the feeding, the turnout
to pasture, the tacking up, the greeting at the gate. She loves
the risk; at most combined driving events someone's cart turns
over. She loves what the horse demands and still teaches:
patience, perseverance — "how you can't muscle a horse and
how you have to rely on better, subtler, nonverbal ways to get
through." One of her horses refused to be harnessed to a cart
for an entire year, but Kumin and Wendy Churchill, a local
trainer hired on for these and other horse-related duties on
the farm, did indeed finally get through.

I watched as Kumin took horse and phaeton (the wooden
cart) for a few warm-up trips around the little exercise track
up on the flat. Then I climbed aboard and joined Churchill,
who continued the workout. Still wearing her white crash helmet, Kumin stood in the center looking on, alert to every
nuance in the horse's step. She has an easy, round laughter that
warms like wool and I heard it often down on the patio; not
so up here: the horse-proud Kumin remained steadfastly and
seriously attentive to the details of training and competing.

It is "a sweet mystery" to her why some individuals seem
immune to the pull of horses; others, so smitten. Kumin had

few friends as a child, and by the age of eight the horse had filled a void in her life. I asked her to ponder the argument, that an obsession with horses marks a retreat from the human race, that to love horses more is to love humans less.

"Then so be it," Kumin said without hesitating. "If I have to sacrifice something to maintain my connection with horses, then let it be. I'm not a hermit, but I don't have a huge need for a lot of social relationships. I'm happy here on the hill, in my solitude."

She once wrote in a poem, "I believe in the gift of the horse, / which is magic . . ." Only by magic, she explained to me, could "something this big, this speedy, this unruly when loose, come to me in this wonderful way. It does seem magic to me to see them running around at liberty, galloping back and forth with their tails over their backs."

Perhaps we make too much of the horse-human connection, she concedes. It may say something of our own foibles, our own desire "to stay connected." This thing for horses — the poet and the professor would agree — is indeed about kinship, about connecting. "When we go out with windfall apples to catch our critters," Kumin wrote in *Women, Animals, and Vegetables* — which reads at times like a primer on horse husbandry — "and they come bucketing in from the far pasture, glistening with good health and high spirits, we know we've caught the right magnificent obsession."

Last stop on my journey was the philosopher's house, but Vicki Hearne — author of *Adam's Task*, *Animal Happiness* and *Bandit*, among others — was ailing that day and we never did meet except on the phone. I had to imagine her house in the Connecticut woods at the end of Horsehill Road, where we were to have coffee by a red barn and talk horses. I have the impression that her house on an acre and a half of land is not entirely rural, but rural enough that her cats must remain indoor cats or be preyed upon by neighborhood coyotes.

I first encountered the name Vicki Hearne in a *New Yorker* article about horses published in the late 1980s and I have been a fan of her writing ever since. As philosophers are wont to do, she sees the complexity of things, and her writing often demands attentiveness. A poet and an animal trainer, she has a deep and abiding interest in the language of animals. A former professor at Yale, she possesses an innate desire to teach, and what I take from her writing is an ongoing theme about the sophistication of animals and the need for humans to pay more attention to catch the creatures' drift.

"My thinking, such as it is," she wrote in *Animal Happiness*, "I learned from the animals, for whom happiness is usually a matter of getting the job done. Clear that fence, fetch in those sheep, move those calves, win that race, find that guy, retrieve that bird." An accomplished rider who has trained dogs and horses for at least three decades, Hearne argues — and I believe she is right — that the happy horse is the one who knows his job and does it.

Hearne's earliest conscious memory is of sitting on a fence reading a story by Rudyard Kipling, with her collie and a horse nearby. It struck her then that dog, horse and poetry were all she really needed in life.

"Our senses sweetly ordered / in the lift and fall of hooves," Hearne once wrote in a poem. I asked her to put that in plainer terms, and she replied — as I knew she would — that she had already *done* so. This much she did say: "When I wrote that poem I was unable to conceive of poetry without conceiving of horsemanship, and vice versa."

We traded a few horse stories, and she told me about her horse Peppy (aka Peppermint Twist), an Appaloosa she once jumped seven feet nine inches at a puissance event in San Francisco. "When I met him," said Hearne, "he was crazy, but he did love to jump. He went from being a killer horse to being my safest school horse. I had a blind student and Peppy knew she was blind and took care with her. If a beginner rider was on his back and was shifting forward, he would do something with his shoulder and helpfully reseat the rider. But if you put someone around him who did not respect him, he'd see horse eaters everywhere. One day there was a man in the paddock next to his. The man was a dude and he was just throwing his weight around. Peppy jumped from his paddock and came within two inches of that man."

"Just to blow some wind at him?" I asked.

"Yes," said Hearne. "Just to blow some wind at him."

Like many others, she seems to have been possessed early by thoughts of horses. She remembers as a toddler being in a car with her father as they drove past a herd of horses and asking what those animals were. "Horses," her father replied. "Where's *my* horse?" young Vicki wanted to know.

Hearne understands the many ways that horses attract humans. *Nostalgia* — "Horsemanship is like any of the humanities. Its impetus is a fiction about restoring the golden age." *Friendship* — "Basically, when I buy a horse or a dog or a cat, he or she has a home with me for life." *Passion* — Our stories about horsemen, she observed in *Adam's Task*, whether from history books or children's tales of horse-inspired heroism, "are allegories about what it is to know what interests you." So many people go through life without ever discovering what it is they care most deeply about; those who love horses have no such doubts, and so it is, says Hearne, that "the passion for a life with horses is so powerful in this culture."

Monty Roberts will tell you that his book, *The Man Who Listens to Horses*, is blessed with a poetic — and quite wrongheaded — title. The horse, especially the wild horse, he will say, is largely a quiet animal. Monty does not so much listen to horses; he watches them with a keen and learned eye. In the same way, the horse does not so much listen to his rider as read him. But in the notion of listening — by both horses and humans — is contained a kind of poetic truth that moves us closer yet to understanding horse fever.

Elizabeth Atwood Lawrence told me about a study done of girls at Pony Clubs. A very high percentage of girls polled actually confided in their horses and ponies. As far as those girls were concerned, their horses did them the courtesy of at least listening. Maxine Kumin believes that what lies at the heart of the girl-horse whirlwind connection is the horse's constancy in a world that seems to the girl fragile and unreliable.

The horsiness of young girls has long been seen as sexual, and the New York City artist Janet Biggs finds at least some wisdom in that explanation. Yet she would also agree with Kumin: the horse offers to young girls a sense of control in a world that seems to them quite out of control.

Biggs has put on so many horse-related art exhibitions and video installations that she now jokingly refers to herself as "a horse artist." In one, called *Girls and Horses* (the photograph "Celeste in Her Bedroom," on page 251, comes from that exhibition), a rotating projector casts on the wall a twelve-foot-tall image of a girl riding a dressage horse, while eight television monitors elsewhere in the gallery show other images: children playing horse with each other; riding mechanical horses, stick ponies, their fathers' backs. The piece, said Biggs, explores childhood fears, anxieties and, maybe most of all, pleasure.

"The attraction that young girls feel for horses," Biggs told me, "is an attraction to power that girls don't experience elsewhere in their lives. You can have no control over other parts of your life, but you get on this twelve-hundred-pound animal and he does what you say. Part of pleasure is power. Beauty is also part of the attraction, and so is historical romance."

Biggs rode as a child and teenager, then left riding for eighteen years until she rediscovered it three years ago. She now rides five days a week, rising at 4:30 A.M. daily to teach riding or train horses or, as an auxiliary parks patroller, to ride in Central Park. Janet Biggs has rediscovered a childhood passion. Some of the ribbons in that photograph "Celeste in Her Bedroom" were won by Janet when she was as young and horse keen as Celeste. One New York critic observed of *Girls and Horses* that "The overriding picture . . . is one of happy girls consumed."

Horses are such wonderful teachers, horse gentlers say. It sounded mystical when I first heard it, even silly. I conjured this horse in a scholar's cap at the blackboard, pointer in hoof, citing a quotation from Hegel . . .

I know a little of what that phrase intends, thanks in part to a letter from another rider on that Wyoming ride.

Gill E. lives on a horse farm in Suffolk. I had asked her why she was so drawn to horses, and she remarked that this was perhaps the first time she had given any serious thought to the matter, so much are horses a given in her life. In the measured phrases of her handwritten letter, the gs and fs and ys curving gently below the line like barbless hooks, Gill reminded me that the horse is indeed a teacher without ever intending to be.

Gill has two daughters, Sara and Hannah. Each had a pony as a child. Pheasant gave Sara the confidence she would need, Pippa taught Hannah patience. If Hannah was rough the pony would abruptly stop, sending her rider over her head into a heap on the ground "and Pippa would graze with a smug expression on her face."

Ponies and horses have punctuated Gill's entire life. As girls, Gill and her sister Ro joined the Pony Club, where they sometimes fed their little mounts chocolate and fizzy lemonade. Gill remembered her first boyfriend and how they rode out together hand in hand. The part-Thoroughbred gelding she competed with as a teenager was eventually sold — to buy a car. Husband Michael led to daughters, then new ponies, and another horsy generation was launched.

When Gill's mother fell prey to cancer, Gill and Ro nursed her at home over the course of two long, hard years. Ro, early in this period, had given Gill two young Arab horses — "a large stroppy gelding called Dan and a worried dainty mare called Spookie." Hannah, Sara, Gill and Ro spent the next year breaking the horses, who sustained them during a dark and difficult time. "Our horses," Gill wrote, "helped us through our grieving. Just to go down to the meadow in the evening to touch and smell Spookie was healing, soothing and very reassuring."

If the horse is magic, as Maxine Kumin believes, the enchantment is not confined to those, like Gill, who have ridden since their youth. Nancy, another woman on that Wyoming ride, reminded me that discovering horses late can be better than coming to them early.

"To some of us the touch of a horse's body, the way he tilts his head toward you, the feel of soft mane in your hands, the smell of his sweet breath and the intoxicating effect of burying your face in his neck are all pure joy," Nancy wrote from rural Vermont. I have watched her with her two Arab horses in the paddock and marveled at her marveling. One of my photographs from Wyoming has her in what might be a trademark pose: a joyful, face-first embrace of the horse's neck.

"I was not one of the fortunate ones," says Nancy, "to grow up with the smell of manure on my shoes and the privilege of riding lessons. I waited to fulfill my dreams till I was forty. Perhaps the years ahead will be all the more precious for me because the gift was so long in coming."

The gift of the horse comes with all sorts of strings. The horse is a huge and potentially dangerous animal who bears the stamp of every human — kind and otherwise — who ever owned or rode him. Obsession with horses can be obsession with power, and many horses know that to be true.

It would be foolhardy to presume that affection alone, or trust or care, would earn every horse's loyalty; or that loyalty, once won, somehow safeguards your bones and those Chiclet teeth that Bill Barich described. My own contact with the horse and with horse-wise people has caused my respect for the horse to grow immensely. I like the word *respect* in any case, but I especially find it appropriate to human-horse relationships. It should land us in a good zone — somewhere between wariness and wonder.

If many of us are still wild about horses, it is because the horse still matters, maybe now more than ever:

Because the horse literally lifts the rider up off the ground and lets her, lets him, see the world in a singular way.

Because the delicate balance in riding between risk and power still delights and rewards.

Because the horse still speaks to us of elegance and beauty, spirit and proud lineage.

Because horses took us to war, plowed our fields, pulled our wagons across continents, and the memory still feels fresh.

Because to get close to a horse is to feel a kinship with the great tribe of horses long gone, with Ruffian and Secretariat, with Comanche and Bucephalus.

Because partnership with a horse is ancient and primal and consuming, and writers and storytellers are still drawn to that territory, so that riding begets reading.

Because the horse offers us, even those of us of peasant stock, the sense of privilege that royalty felt.

Because there is no promise like the promise of a foal, no journey like one on horseback, no sight so pleasing as horses grazing in a paddock, no thought so warming as the knowledge that free horses still run on the plains.

Because to sit astride a walking horse is to banish time and to live, as the horse lives, in the moment.

Because no other animal lets you partake so directly, so sensuously, in what a fleet horse feels — the tickle of wind, the kettledrumming of hooves, the easy grace of the trot.

And because there is nothing quite like a gallop in wide-open spaces, when human and horse and the earth below feel all of a piece, when heartbeat and hoofbeat find each other's rhythm.

FURTHER READING

INTRODUCTION

Evans, Nicholas. *The Horse Whisperer*. New York: Dell Publishing, 1996.

Masson, Jeffrey Moussaieff and Susan McCarthy. *When Elephants Weep: The Emotional Lives of Animals*. New York: Delta Publishing, 1996.

McCarthy, Cormac. *All the Pretty Horses*. New York: Vintage Books, 1993.

———. *The Crossing*. New York: Alfred A. Knopf, 1994.

Millar, Ian and Larry Scanlan. *Riding High: Ian Millar's World of Show Jumping*. Toronto: McClelland & Stewart, 1990.

Roberts, Monty. *The Man Who Listens to Horses*. New York: Random House, 1997.

Scanlan, Lawrence. *Big Ben*. Toronto: Scholastic, 1994.

———. "Flying High." *Equinox* 69, June 1993, 34–45.

———. "Why Humans Love Horses." *Equinox* 85, February 1996, 24–35.

Chapter 1

HEAVENLY HORSES

Anthony, David, Dimitro Y. Telegin and Dorcas Brown. "The Origin of Horseback Riding." *Scientific American*, December 1991, 94–100.

Barclay, Harold B. *The Role of the Horse in Man's Culture*. London: J. A. Allen, 1980.

Clark, Ella E. *Indian Legends from the Northern Rockies*. Norman, Okla.: University of Oklahoma Press, 1966.

Clark, La Verne Harrell. *They Sang for Horses: The Impact of the Horse on Navajo and Apache Folklore*. Tucson, Ariz.: University of Arizona Press, 1966.

Denhardt, Robert M. *The Horse of the Americas*. Norman, Okla.: University of Oklahoma Press, 1975.

de Vries, Ad. *Dictionary of Symbols and Imagery*. Amsterdam: North Holland, 1974.

Dobie, J. Frank. *On the Open Range*. Dallas: Banks Upshaw, 1940.

Ehrlich, Gretel. *The Solace of Open Spaces*. New York: Penguin USA, 1985.

Gregg, Josiah. *Commerce of the Prairies*. New York: Readex Books, 1966.

Howey, Oldfield M. *The Horse in Magic and Myth*. New York: Castle Publishing, 1958.

Jankovich, Mikos. *They Rode Into Europe: The Fruitful Exchange in the Arts of Horsemanship Between East and West*. Translated by Anthony Dent. London: George Harrap & Co., 1968.

Jurmain, Suzanne. *Once Upon a Horse: A History of Horses — And How They Shaped Our History*. New York: Lee & Shepard Books, 1989.

Lame Deer, John (Fire). *Lame Deer, Seeker of Visions: The Life of a Sioux Medicine Man*. New York: Simon and Schuster, 1972.

Law, Robin. *The Horse in West African History*. London: Oxford University Press, 1980.

Lawrence, Elizabeth Atwood. *Hoofbeats and Society: Studies of Human-Horse Interactions*. Bloomington, Ind.: Indiana University Press, 1985.

Leakey, Richard. *The Making of Mankind*. New York: Dutton, 1981.

Norman, Philip. "All the King's Horses." *Sunday Times Magazine*, 7 December 1997, 44–50.

Plutarch. *The Age of Alexander*. New York: Penguin USA, 1980.

Roe, Frank G. *The Indian and the Horse*. Norman, Okla.: University of Oklahoma Press, 1955.

Shepard, Paul. *The Others: How Animals Made Us Human.* Washington, D.C.: Island Press, 1996.

Steele, Rufus. *Mustangs of the Mesas.* Hollywood: Murray and Gee, 1941.

Uden, Grant. *High Horses.* Harmondsworth, England: Kestrel Books, 1976.

Vernam, Glenn. *Man on Horseback.* New York: Harper & Row, 1964.

Chapter 2

WILD ABOUT WILD HORSES

Berger, Joel. *Wild Horses of the Great Basin.* Chicago: The University of Chicago Press, 1986.

Birdsell, Sandra. *The Two-Headed Calf.* Toronto: McClelland & Stewart, 1997.

Bower, Joe. "Planned Parenthood." *Audubon* 97, July–August 1995, 20.

Collins, Jerry. "Mange on the Range: The Chilcotin's Wild Horses Are Judged to Be a Nuisance." *Western Report*, 1 May 1995, 14.

Curtin, Sharon, Yva Momatiuk and John Eastcott. *Mustang.* Bearsville, N.Y.: Rufus Publications, 1996.

Dary, David. *Cowboy Culture: A Saga of Five Centuries.* New York: Avon Books, 1982.

Dean, Cornelia. "Horses of Coast Islands, a Regional Symbol, Harm the Environment." *New York Times*, 27 July 1993, C4.

Dobie, J. Frank. *The Mustangs.* London: Hammond, Hammond & Co., 1954.

Dolphin, Ric and Marilyn McKinley. "A Death Threat to the Prairie Ponies." *Alberta Report*, 22 March 1982, 36–40.

Edwards, Elwyn Hartley, ed. *Encyclopedia of the Horse.* London: Peerage Books, 1985.

———. *The Ultimate Horse Book.* New York: Dorling Kindersley, 1991.

————. *Wild Horses: A Spirit Unbroken*. Stillwater, Minn.: Voyageur Press, 1995.

Fazio, Patricia Mabee. "The Fight to Save a Memory: Creation of the Pryor Mountain Wild Horse Range (1968) and Evolving Federal Wild Horse Protection." Ph.D. thesis, Texas A&M University, 1995.

Flaherty, Kathleen. *On the Skyline Trail in Jasper National Park*. Ideas, CBC Radio Transcripts, 23 March 1995.

Green, Ben K. *A Thousand Miles of Mustangin'*. Flagstaff, Ariz.: Northland Press, 1972.

Hall, E. T. *The Dance of Life*. Garden City, NY: Doubleday, 1983.

Harbury, Martin. *The Last of the Wild Horses*. Garden City, NY: Doubleday & Co., 1984.

Irving, Washington. *A Tour on the Prairies*. Norman, Okla.: University of Oklahoma Press, 1971.

Keough, Pat and Rosemarie. *Wild and Beautiful Sable Island*. Fulford Harbour, BC: Nahanni, 1993.

Kirkpatrick, Jay. F. *Into the Wind: Wild Horses of North America*. Photography by Michael H. Francis. Minocqua, Wis.: Northword Press, 1994.

Klinkenborg, Verlyn. "The Mustang Myth." *Audubon*, January–February 1994, 36–51.

McInnis, Doug. "Hold Those Horses." *Montreal Gazette*, 31 March 1996, B4.

Nagle, Patrick. "Wild Horses Raise Prairie Dust Storm." *Toronto Star*, 3 July 1993, D6.

Ryden, Hope. *America's Last Wild Horses*. New York: E. P. Dutton & Co., 1970.

————. *Mustangs: A Return to the Wild*. New York: Viking Press, 1972.

Scott, Jack Denton and Ozzie Sweet. *Island of Wild Horses*. New York: Putnam, 1978.

Spragg, Mark, ed. *Thunder of the Mustangs: Legend and Lore of the Wild Horse*. San Francisco: Sierra Club, 1997.

Wyman, Walker D. *The Wild Horse of the West*. Lincoln, Nebr.: University of Nebraska Press, 1965.

Chapter 3

THE HORSE THROUGH THE
LOOKING GLASS

Anglesey, The Marquess of. *A History of the British Cavalry, 1816 to 1919*, vol. II, *1851 to 1871*. London: Secker & Warburg, 1982.

Budiansky, Stephen. *The Nature of Horses: Exploring Equine Evolution, Intelligence, and Behaviour*. New York: The Free Press, 1997.

Clifton, Merritt and The Animal Legal Defense Fund. "Kitty Cruelty: Why Is Animal Abuse Punished So Lightly?" *Utne Reader*, January/February 1993: 131–132.

Gould, Stephen Jay. *Dinosaur in a Haystack: Reflections in Natural History*. New York: Harmony Books, 1995.

Gzowski, Peter. *An Unbroken Line*. Toronto: McClelland & Stewart, 1984.

Hyland, Ann. *Equus: The Horse in the Roman World*. New Haven: Yale University Press, 1990.

Lowie, Robert. *The Indians of the Plains*. Garden City, NY: Natural History Press, 1963.

MacEwan, Grant. *Memory Meadows: Horse Stories from Canada's Past*. Vancouver: Greystone, 1997.

Morris, Desmond. *Horsewatching*. New York: Crown, 1989.

Noyes, Stanley. *Los Comanches: The Horse People*. Albuquerque, NM: University of New Mexico Press, 1993.

Swift, Jonathan. *Gulliver's Travels*. New York: Penguin USA, 1989.

Trench, Charles Chenevix. *A History of Horsemanship*. Garden City, NY: Doubleday, 1970.

Chapter 4

THE GENTLE ART OF THE
HORSE WHISPERER

Ainslie, Tom and Bonnie Ledbetter. *The Body Language of Horses*. New York: William Morrow & Co., 1980.

Anderson, J. K. *Ancient Greek Horsemanship*. Berkeley: University of California Press, 1961.

Blundeville, Thomas. *The Arte of Ryding and Breakinge Greate Horses*. New York: Da Capo Press, 1969.

Davis, Susan. "Gentle Hands for Wild Horses." *Sports Illustrated* 85, no. 20, 11 November 1996, 7–9.

Dorrance, Tom. *True Unity: Willing Communication Between Horse and Human*. Tuscarora, Nev.: Give-It-A-Go, 1996.

Fredriksson, Kristine. *American Rodeo: From Buffalo Bill to Big Business*. College Station, Tex.: Texas A&M University Press, 1985.

Hunt, Ray. *Think Harmony With Horses*. Tuscarora, Nev.: Give-It-A-Go, 1991.

Johnson, Dirk. "Broncobusters Try New Tack: Tenderness." *New York Times*, 11 October 1993, A1, A13.

Littauer, Vladimir S. *The Development of Modern Riding: The Story of Formal Riding from Renaissance Times to the Present*. New York: Howell Book House, 1991.

Lorenz, Konrad Z. *King Solomon's Ring: New Light on Animal Ways*. London: Methuen, 1952.

McGovern, Celeste. "A Kinder, Gentler Cowboy Shows How." *Alberta Report*, 21 November 1994, 33–34.

Morris, George H. *Hunter Seat Equitation*. New York: Doubleday, 1979.

Newman, Judith and Laird Harrison. "Horse Sense." *People*, 6 December 1993, 51–52.

Rarey, J. S. *The Farmer's Friend, Containing Rarey's Horse Secret, With Other Valuable Receipts and Information*. Hamilton, C.W.: Porter and Schneider, 1858.

———. *The Modern Art of Taming Wild Horses*. Bedford, Mass.: Applewood Books, 1995.

Self, Margaret Cabell. *Horsemastership: Methods of Training the Horse and the Rider*. New York: A. S. Barnes & Co., 1952.

van der Post, Laurens. *About Blady: A Pattern Out of Time*. London: Chatto & Windus, 1991.

Xenophon. *The Art of Horsemanship*. Boston: Little Brown & Co., 1893.

Chapter 5

THE HORSE IN BATTLE

Jordens, Lorraine. "Horses and Mules: The Forgotten Soldiers."
     Alberta History (Spring 1993): 20–26.

Keegan, John. *A History of Warfare*. Toronto: Vintage Books, 1993.

Kelly, William and Nora. *The Horses of the Royal Canadian
     Mounted Police: A Pictorial History*. Toronto: Doubleday &
     Co., 1984.

Ketelsen, James. "Mall Mounties." *Forbes*, 17 June 1996, 84–85.

Lamb, A. J. R. *The Story of the Horse*. London: Alexander
     Maclehose & Co., 1938.

Lawrence, Elizabeth Atwood. *His Very Silence Speaks: The Horse
     Who Survived Custer's Last Stand*. Detroit: Wayne State
     University Press, 1989.

Seely, General Jack. *My Horse Warrior*. London: Hodder &
     Stoughton, 1934.

Spuler, Bertold. *History of the Mongols: Based on Eastern and
     Western Accounts of the Thirteenth and Fourteenth Centuries*.
     Berkeley: University of California Press, 1972.

Stoneridge, M. A. *Great Horses of Our Time*. New York:
     Doubleday & Co., 1972.

Tamblyn, Lt.-Col. D. S. *The Horse in War and Famous Canadian
     War Horses*. Kingston, Ont.: The Jackson Press, 1931.

Wallace, Ernest and E. Hoebel, *The Comanches*. Norman, Okla.:
     University of Oklahoma Press, 1952.

Chapter 6

THE WONDER HORSES OF
HOLLYWOOD AND LITERATURE

Baldwin, Neil. *Inventing the Century*. New York: Hyperion, 1995.

Baxter, John. *Stunt: The Story of the Great Movie Stunt Men*.
     London: Macdonald, 1973.

Blake, Henry. *Talking With Horses*. New York: E. P. Dutton, 1976.

Cary, Diana Serra. *The Hollywood Posse: The Story of a Gallant Band of Horsemen Who Made Movie History*. Boston: Houghton Mifflin, 1975.

Fox, Charles Philip. *A Pictorial History of Performing Horses*. New York: Bramhall House, 1960.

Golden, Lilly, ed. *The Literary Horse: Great Modern Stories About Horses*. New York: Atlantic Monthly Press, 1995.

Henry, Marguerite. *Black Gold*. Chicago: Rand McNally, 1957.

Hintz, H. F. *Horses in the Movies*. New York: A. S. Barnes, 1979.

James, Will. *Smoky the Cow Horse*. New York: Charles Scribner's Sons, 1926.

Macauley, Thurston, ed. *The Great Horse Omnibus: From Homer to Hemingway*. New York: Ziff-Davis, 1949.

McCourt, Frank. *Angela's Ashes: A Memoir*. New York: Scribners, 1996.

Muybridge, Eadweard. *Animals in Motion*. New York: Dover Publications, 1957.

O'Hara, Mary. *Flicka's Friend: The Autobiography of Mary O'Hara*. New York: Putnam, 1982.

———. *My Friend Flicka*. Philadelphia: J. B. Lippincott Company, 1941.

Rothel, David. *The Singing Cowboys*. New York: A. S. Barnes, 1978.

Rudolph, Alan and Robert Altman. *Buffalo Bill and the Indians, or Sitting Bull's History Lesson*. New York: Bantam, 1976.

Sewell, Anna. *Black Beauty*. Middlesex, England: Penguin, 1972.

Simpson, George Gaylord. *Horses: The Story of the Horse Family in the Modern World and Through Sixty Million Years of History*. New York: Oxford University Press: 1961.

Spoto, Donald. *A Passion for Life: The Biography of Elizabeth Taylor*. New York: HarperCollins, 1995.

Stott, Greg. "Wanted: Will James." *Equinox*, 80, March 1995, 55–65.

Tuska, Jon. *The Filming of the West*. New York: Doubleday & Co., 1976.

Urquhart, Fred. *The Book of Horses*. London: Secker & Warburg, 1981.

Vanderhaeghe, Guy. *The Englishman's Boy*. Toronto: McClelland & Stewart, 1996.

Wise, Arthur and Derek Ware. *Stunting in the Cinema*. London: Constable, 1973.

Chapter 7

SPORT HORSE LEGENDS

Blum, Howard. "The Horse Murders." *Vanity Fair*, January 1995, 92–101, 138–140.

Brown, Robin. *The Inside Track*. CBC Radio, 20 October 1996.

Cauz, Louis E. *The Plate: A Royal Tradition*. Toronto: Deneau, 1984.

Davis, J. Madison. *Dick Francis*. Boston, Mass.: Twayne Publishers, 1989.

Edwards, Elwyn Hartley and Candida Geddes. *The Complete Book of the Horse*. Edmonton, Al.: Hurtig, 1982.

Englade, Ken. *Hot Blood: The Money, the Brach Heiress, the Horse Murders*. New York: St. Martin's Press, 1996.

Klimo, Kate. *Heroic Horses and Their Riders*. New York: Platt and Munk, 1974.

Lidz, Franz. "He's No Paperback Rider." *Sports Illustrated* 79, no. 20, 15 November 1993, 106, 108.

Menino, Holly. *Forward Motion: Horses, Humans, and the Competitive Enterprise*. New York: North Point Press, 1996.

Nack, William. "Pure Heart." *Sports Illustrated* 81, no. 17, 24 October 1994, 76–88.

Nack, William and Lester Munson. "Blood Money." *Sports Illustrated* 77, no. 21, 16 November 1992, 18–28.

Robertson, William H. P. *The History of Thoroughbred Racing in America*. Inglewood Cliffs, N.J.: Prentice Hall, 1964.

Schefstad, Anthony J. "The Backstretch: Some Call It Home." Ph.D. thesis, University of Maryland, 1995.

Smith, Gene. "Ruffian." *American Heritage*, September 1993, 46–57.

Stegner, Wallace. *Collected Stories of Wallace Stegner*. New York: Wings Books, 1994.

Turcotte, Ron. "Secretariat and Me." *Sports Illustrated* 78, no. 17
    3 May 1993, 45–49.
Wilkinson, Michael. *The Phar Lap Story*. Dingley, Victoria,
    Australia: Budget Books, 1983.

Chapter 8

EPIC RIDES

Burnaby, Frederick Gustavus. *A Ride to Khiva*. London: Cassell
    Petter & Galpin, 1877.
Clébert, Jean Paul. *The Gypsies*. New York: Dutton, 1963.
Dodwell, Christina. *A Traveller on Horseback*. London: Hodder &
    Stoughton, 1987.
Granfield, Linda. *Cowboy: An Album*. New York: Ticknor & Fields
    Books, 1994.
Kelly, Charles. *The Outlaw Trail: A History of Butch Cassidy and His
    Wild Bunch*. Lincoln, Nebr.: University of Nebraska Press, 1996.
Parkman, Francis. *The Oregon Trail*. Edited by E.N. Feltskog,
    Madison, Wisc.: University of Wisconsin Press, 1969.
Tschiffely, Aime Felix. *Tschiffely's Ride: Ten Thousand Miles in the
    Saddle from Southern Cross to Pole Star*. New York: Simon &
    Schuster, 1933.
Ulyatt, Kenneth. *The Day of the Cowboy*. Middlesex, England:
    Longman Young Books, 1973.
Walker, Eric. *The Great Trek*. London: Black, 1965.
Whittome, Barbara. *Russian Ride: The Account of a 2,500 Mile
    Trek with Three Cossack Horses*. London: Boxtree, 1996.

Chapter 9

MY KINGDOM FOR A — PONY

Ewers, John. *The Horse in Blackfoot Indian Culture*. Bulletin 159.
    Washington D.C.: Smithsonian Institution, Bureau of American
    Ethnology, 1955.

Settle, Raymond W. and Mary Lund Settle. *Saddles and Spurs: The Pony Express Saga*. Lincoln, Nebr.: University of Nebraska Press, 1955.

Siringo, Charles A. *A Texas Cowboy, or 15 Years on the Hurricane Deck of a Spanish Pony*. New York: Umbedenstock, 1950.

Thomas, Heather Smith. *The Wild Horse Controversy*. South Brunswick, N.J.: A. S. Barnes, 1979.

Webber, Toni. *The Pony-Lover's Handbook*. London: Pelham, 1973.

Chapter 10

HORSE TALES TALL AND TRUE

Camp, Charles. *Muggins, the Cow Horse*. Denver: Welsh-Haffner Printing, 1928.

Dobie, J. Frank, C. Boatright Moady, and Harry Ransom, editors. *Mustangs and Cow Horses*. Dallas: Southern Methodist University Press, 1965.

Godwin, Don and Vi. *Faith vs. Fear*. Mulvane, Kans.: Donald and Vevia C. Godwin, 1985.

Goehner, Amy Lennard. "Animal Magnetism." *Sports Illustrated* 80, no. 7, 21 February 1994, 84–85.

Kendall, George Wilkins. *Narrative of the Texas Santa Fe Expedition*. Chicago: Lakeside, 1929.

Kinnish, Mary Kay, ed. *Understanding Equine Behavior: Know What's On Your Horse's Mind*. Gaithersburg, Md.: Fleet Street Publishing, 1996.

Maeterlinck, Maurice. *The Unknown Guest*. New York: Dodd, Mead and Company, 1914.

Pron, Nicholas. "Horse Falls From Truck on Busy 401." *Toronto Star*, 30 November 1988, A7.

EPILOGUE

Barich, Bill. *Laughing in the Hills*. New York: Penguin USA, 1981.

Bixby-Hammett, Doris and William H. Brooks, "Common Injuries

in Horseback Riding: A Review." *Sports Medicine* 9, no. 6
(1990): 36–47.

Hearne, Vicki. *Adam's Task: Calling Animals by Name*. New York:
HarperPerennial, 1994.

———. *Animal Happiness*. New York: HarperCollins, 1994.

———. *Bandit: Dossier of a Dangerous Dog*. New York:
HarperPerennial, 1991.

Kumin, Maxine. *Connecting the Dots*. New York: W. W. Norton,
1996.

———. *In Deep: Country Essays*. New York: Viking, 1987.

———. *Looking for Luck*. New York: W. W. Norton, 1992.

———. *Women, Animals, and Vegetables: Essays and Stories*. New
York: W. W. Norton, 1994.

Lang, Gerald and Lee Marks. *The Horse: Photographic Images,
1839 to the Present*. With an essay by Lawrence, Elizabeth
Atwood. New York: Harry N. Abrams, 1991.

Lawrence, Elizabeth Atwood. *Rodeo: An Anthropologist Looks at
the Wild and the Tame*. Chicago: University of Chicago Press,
1984.

———. "Actor's Injury Doesn't Dissuade Riders in
Horse Country." *New York Times*, 4 June 1995, 34.

IMAGE CREDITS

[photo credit for page vi] Photograph by Astrid Palmowski.

1.1 Reprinted with permission of the Historical-Archaeological Experimental Centre, Lejre, Denmark.

1.2 © Alen Mac Weeney. White Pony, Clifden Horse Fair, Ireland, 1965.

1.3 Drawing by Walter Crane. From the Corbis-Bettmann Archive, New York.

1.4 Artist unknown. From SuperStock.

1.5 *The Four Horsemen of the Apocalypse*, by Albrecht Dürer. From the Corbis-Bettmann Archive.

2.1 © John Eastcott/Yva Momatiuk.

2.2 © David Hurn, Magnum Photos.

2.3 © Rita Summers.

2.4 © John Eastcott/Yva Momatiuk.

2.5 © John Eastcott/Yva Momatiuk.

3.1 Photographer unknown. From Circus World Museum, Baraboo, Wisconsin.

3.2 Lascaux Cave Painting, Chinese Horse. The Corbis-Bettmann Archive.

3.3 © Gary Leppart.

3.4 *Sun River War Party*, Charles Marion Russell, oil on canvas, 1903. Rockwell Museum, Corning, New York.

3.5 *Akbar Hunting a Tiger*. SuperStock.

4.1 *One of the Rough String*. Charles Marion Russell, 1913, oil on canvas, from the Glenbow Collection, Calgary, Alberta, Canada.

4.2 Cowgirl at the first Calgary Stampede, 1912. From the Glenbow Archives, Calgary, Alberta, Canada.

4.3 © Monty Roberts. Photograph by Christopher Dydyk.

4.4 Reprinted with the permission of *Alberta Report*.

4.5 From the Glenbow Archives, Calgary, Alberta, Canada.

5.1 *Scotland Forever! The Charge of the Scots Greys at Waterloo, 1881*, oil on canvas. By Lady Elizabeth Butler. Reprinted with permission of Leeds Museum and Galleries (City Art Gallery). Photograph by Courtauld Institute of Art, London, England.

5.2 *Mongol Archer*, Ming Dynasty drawing. SuperStock.

5.3 *Female Knight on Black Horse*, Capodilista Codex manuscripts, Biblioteca Civica, Padua, Italy. SuperStock.

5.4 Comanche, survivor of Custer's massacre. Corbis-Bettmann Archive, New York.

5.5 "Good-bye, Old Man," 1916. By Fortunino Matania. Reprinted by permission of The Blue Cross (London, England) and Canadian War Museum (Ottawa).

6.1 Nicholas Konrad, age seven, Gladstone, Manitoba, Canada, 1928. Courtesy of Joe Konrad.

6.2 The Kobal Collection, New York, N.Y.

6.3 The Kobal Collection.

6.4 Eadweard Muybridge. Plate 640 from *Animals in Motion*, 1887. Reprinted with permission of The Minneapolis Institute of Arts.

6.5 Corbis-Bettmann Archive.

7.1 © Shawn Hamilton.

7.2 Portrait of Ruffian by Richard Stone Reeves, National Museum of Racing and Hall of Fame, Saratoga Springs, New York.

7.3 Corbis-Bettmann Archive.

7.4 Corbis-Bettmann Archive.

7.5 © Todd Korol.

8.1 *In Without Knocking*, by Charles Marion Russell, oil on canvas, 1909; #1961.201. From the Amon Carter Museum, Fort Worth, Texas.

8.2 The photographer Erwin E. Smith stopping at the chuck wagon for a cup of coffee. LS Ranch, Texas. 1908; #LC.S59.133. From the Erwin E. Smith Collection of the Library of Congress on deposit at the Amon Carter Museum, Fort Worth, Texas.

8.3 Tschiffely and Mancha greeted by New York mayor James Walker in 1928. Corbis-Bettmann Archive.

8.4 Photograph courtesy Barbara Whittome.

8.5 "Romania, 1968." © Josef Koudelka, Magnum Photos.

9.1 *Celeste in Her Bedroom*. © Janet Biggs, 1996. Photo credit: Erma Estwick, courtesy of Anna Kustera Gallery, New York City.

9.2 © Shawn Hamilton.

9.3 *Shoeing the Polo Pony*, by Chloe Henderson, 1905. SuperStock.

9.4 Pony Express Rider Leaving Station. Corbis-Bettmann Archive.

9.5 Buffalo Bill Cody novel cover illustration, 1912. Corbis-Bettmann Archive.

10.1 © *Time* Inc. Ray Bill, photographer.

10.2 *Horse Dives from 75-foot Tower with Rider*, Atlantic City, New Jersey, August 24, 1960. Corbis-Bettmann.

10.3 Reprinted with permission of Holt, Rinehart and Winston, Inc. from *Clever Hans: The Horse of Mr. von Osten*, by Oskar Pfungst, 1965.

10.4 Milk Wagon, Berlin, August 3, 1909. SuperStock.

10.5 © Shawn Hamilton

INDEX